Hugh Macmillan

First forms of vegetation

Hugh Macmillan

First forms of vegetation

ISBN/EAN: 9783337374600

Printed in Europe, USA, Canada, Australia, Japan

Cover: Foto ©berggeist007 / pixelio.de

More available books at **www.hansebooks.com**

PREFACE TO NEW EDITION.

THE first edition of this book was published in 1861, under the name of 'Footnotes from the Page of Nature; or First Forms of Vegetation,' and has been a considerable time out of print. The progress of Cryptogamic Botany during the interval has been such as to necessitate very extensive revision, especially of the last chapter. Having little time for such a task, I would, had I consulted my own inclination, have allowed the book to disappear; believing that it had served its purpose, and satisfied with the measure of success it had achieved. But I have been urged by friends, whose scientific attainments lend weight to their opinion, to re-issue the book with the required additions and corrections—on the ground that, notwithstanding the recent multiplication of separate monographs upon the lowest orders of plant-life, there is still ample room for a comprehensive and popular treatise upon the whole

b

subject. Influenced in this way, I have gone over the volume very carefully, and brought it up as far as I could to the present state of knowledge regarding the topics upon which it treats. Several striking and novel facts connected with all the departments, and especially with fungi, I have been compelled for want of space to omit altogether; while I have been led by the popular nature of the book to glance briefly and superficially over subjects which are worthy of the profoundest study; and upon which, in other circumstances, I should have liked to dwell at more adequate length. I believe, however, that nothing of any real importance has been left out. It is confessedly a difficult task to make such obscure subjects intelligible and interesting to the ordinary reader without sacrificing scientific accuracy. Whether I have succeeded in doing so it is not for me to say; but I am sure that those who best know the difficulty will be most ready to sympathize with my attempt, and to generously overlook all its imperfections, of which none can be more sensible than myself.

Upwards of a hundred pages of new matter have been added; and eleven new illustrations have been embodied in the last chapter, for which I am indebted to the kindness of Mr. Cooke, one of the ablest of our British mycologists. They have been taken from his most admirable ' Handbook of British Fungi,' which no botanist should be without; and were drawn, I believe, by Mr. Worthington Smith, who has done so much by his learned pen and graceful pencil to elucidate the history of these obscure productions. The borrowed wood-cuts, I may mention, are Figs. 27, 28, 31, 32, 33, 34, 35, 40, 43, 48, and 52; all the rest are original.

One of the greatest annoyances of the Cryptogamic student is the constant changing of the nomenclature. More accurate knowledge leading to an improved method of classification has doubtless in many instances been the cause of this procedure. But it is to be feared that some authors introduce innovations capriciously and without any adequate necessity; thus adding greatly to the toil of identification, and to the chaos of synonymy with which our text-books are burdened. On this point I have exercised what I believe to be a wise discretion. I have followed the new fashion to a limited extent, but have retained a good many old names owing to their own fitness and the precious associations that have gathered around them by the study of years. By other new-fangled names these lowly plants would not have smelt so sweet to me.

As this edition is virtually a new book, I have discarded the former principal title on account of its obscurity and fancifulness, and adopted the alternative title as the sole one, believing that it recommends itself by its simplicity and comprehensive descriptiveness.

I may add, in the words of the original Preface, that the different chapters of this book were first composed and delivered in the form of a series of popular lectures. This circumstance will account for their style and general tone, which have been preserved although re-written and considerably extended, and which may be supposed to be better adapted for listeners than readers. I had serious thoughts of re-casting the whole work on this occasion, and presenting it in a more systematic mould. But this would require too much labour; and it is doubtful if the gain after all would have been a sufficient compensation. My object in publishing the book is not so much

to impart cut-and-dried information as to kindle
the sympathy and awaken the interest of the
reader in a department of nature with which few,
owing to the technical phraseology. of botanical
works, are familiar. Those who have derived
pleasure and profit from the study of flowers
and ferns—subjects, it is pleasing to find, now
everywhere popular—by descending lower into
the arcana of the vegetable kingdom will find a
still more interesting and delightful field of research in the objects brought under review in the
following pages. This work is neither a text-book
nor a guide to species, but simply a popular history
of the uses, structural peculiarities, associations,
and other interesting facts connected with the
humblest forms of plant-life ; and, as such, it may
be regarded as an introduction to more scientific
treatises which deal with particular orders and
species.

H. M.

June 1874.

CONTENTS.

INTRODUCTION.

IT cannot have escaped the notice even of the most unobservant, that the tendency to vegetate is a power restless and perpetual. It has been in operation from the earliest ages of the earth, ever since living beings were capable of existing upon its surface, as evinced by the fossil remains found in the most ancient rocks. Like a palimpsest, the successive strata of the earth have been covered with successive races of plants. Wherever an igneous rock was upheaved into the sky by some internal convulsion, its bare sides and summit were speedily covered with vegetation ; wherever the water retired, leaving its sediment behind, the dry land thus formed became, in a wonderfully short space of time, clothed with verdure. From pole to pole, each stratum of soil, as soon as deposited, was adorned with a rich exuberance of plant-life. Nor is the layer of Nature's floral handwriting which

A

now appears on the surface less extensive, as compared with the page, than the buried and partially
obliterated layers beneath, though the characters be
less grand and imposing.   The earth has lost much
of its primeval fire, and has toned down the rank
luxuriance of its green and umbrageous youth ; but
it still retains a considerable portion of the vigour
which characterized it during the first great period
of organized being—the period distinctively of
herbs and trees.   The whole face of the earth, and
almost every object which belongs to it, is still
strangely instinct with vegetable life.   Coeval in
its origin, it is everywhere present with its indispensable conditions.   Burn down the forest, or
plough the meadow, and from the new soil thus
exposed springs up spontaneously a new crop of
vegetation.   Hew a stone from a quarry, and place
it in a damp situation, and shortly a green tint
begins to creep over it.   Construct a fence of
wooden rails round an enclosure, and in a few
months it is covered with a thin film of primitive
plants.   Expose a pot of jam, or a piece of bread,
or any decayed vegetable or animal matter, to
the air, and in a day or two it will be hoary with
the grey stalks and powdery fructification of the
common mould.   Dam up a stream or the outlet
of a lake, and convert it into a stagnant pond, and
in a few weeks its sides and bottom are covered

with a luxuriant growth of green confervæ, which go on increasing until the water is choked up with vegetable matter, and becomes converted into a bog. How rapidly does Nature bring back into her own bosom the ruin which man has forsaken, harmonizing its haggard features with the softer hues and forms of the scenery around! How quickly does the newly-built wall, which offends the eye by its garishness, become, by the living garniture of mosses and lichens that creep over it, a picturesque object in the landscape! Nature, faithful to her own law, 'Be fruitful, and multiply, and replenish the earth,' produces even in the hottest thermal springs special vegetable forms, whose structure is wonderfully adapted to their situation and requirements, and crimsons even the cold and barren surface of the arctic or alpine snow with a portentous vegetation. She develops a strange weird frost of organic life upon the bark of trees, upon naked rocks, upon the roofs of houses, upon dead and living animal sub-stances, upon glass when not constantly kept clean, and even on iron which had been sub-jected to a red heat a short time before. As if there were not room enough for the amazing profusion of plant-life, she crowds her productions upon each other into the smallest compass, and makes the highest forms the supporters of the lowest. Every inch of ground, however ungenial its climate or unfavourable its conditions, is made available ; every object, however unlikely at first sight, is pressed into her service, and made to bear its burden of life. From the deepest recesses of the earth to which the air can penetrate, to the summits of the loftiest mountains; from the almost unfathomable depths of the ocean to the highest clouds ; from pole to pole, the vast stratum of vegetable life extends ; while it ranges from a temperature of 35° to 135° Fah., a range embracing almost every variety of conditions and circumstances ; and thus the grandly wild Platonic myth of the *cosmos*, as one vast living thing, is not altogether without foundation.

The most cursory and superficial glance will recognise in every scene a class of plants whose singular appearances, habits, and modes of growth are so widely different from those of the trees and flowers around, that they might seem hardly entitled to a place in the vegetable kingdom at all. On walls by the wayside, on rocks on the hills, and on trees in the woods, we see tiny green tufts and grey stains, or particoloured rosettes spreading themselves, easily dried by the heat of the sun, and easily revived by the rain. In almost every stream, lake, ditch, or any collection of standing or moving water, we observe a green slimy matter forming a scum on the surface, or floating in long filaments in the depths. On almost every fallen leaf and decayed branch, fleshy gelatinous bodies of different forms and sizes meet our eye. Sometimes all these different objects appear growing on the same substance. If we examine a fallen, partially decayed twig, half-buried in the earth in a wood, we may find it completely covered with various representatives of these different vegetable growths ; and nothing surely can give us a more striking proof of the universal diffusion of life. All these different plants belong to the second great division of the vegetable kingdom, to which the name of crypto-gamia has been given, on account of the absence, in all the members, of those prominent floral organs which are essential to the production of perfect seed. They are propagated by little embryo plants called spores or sporules, generally invisible to the naked eye, and differing from true seeds in germinating from any part of their surface instead of from two invariable points. Besides this grand distinguishing mark, these plants possess several other peculiar qualities in common. They consist of cells only, and hence are often called cellular plants, in contradistinction to those plants which are possessed of fibres and woody tissue. Their development is also superficial, growth taking

place from the various terminal points ; and hence
they are called acrogens and thallogens, to distinguish them from monocotyledonous and dicoty-
ledonous plants. Popularly, they are known as
mosses, lichens, algæ, and fungi. They open up a
vast field of physiological research. They constitute a microcosm, an *imperium in imperio*, a
strange minute world underlying this great world
of sense and sight, which, though unseen and un-
heeded by man, is yet ever in full and active
operation around us. It is pleasant to turn aside
for a while from the busy human world, with its
ceaseless anxieties, sorrows, and labours, to avert
our gaze from the splendours of forest and garden,
from the visible display of green foliage and
rainbow-coloured blossoms around us, and con-
template the silent and wonderful economy of that
other world of minute or invisible vegetation with
which we are so mysteriously related, though we
know it not. There is something exceedingly
interesting in tracing Nature to her ultimate and
simplest forms. The mind of man has a natural
craving for the infinite. It delights to speculate
either on the vast or the minute ; and we are not
surprised at the paradoxical remark of Linnæus,
that Nature appeared to him greatest in her least
productions.

These plants once occupied the foremost position

in the economy of nature. Like many decayed families whose founders were kings and mighty heroes, but whose descendants are beggars, they were once the aristocracy of the vegetable kingdom, though now reduced to the lowest ranks, and considered the *canaille* of vegetation. Geology reveals to us the extraordinary fact, that one whole volume of the earth's stony book is filled almost exclusively with their history. Life may have been ushered upon our globe through oceans of the lowest types of confervæ, long previous to the deposit of the oldest palæozoic rocks as known to us ; and for myriads of ages these extremely simple and minute plants may have represented the only idea of life on the earth. But passing from conjecture to the domain of established truth, we know of a certainty that at least throughout the vast periods of the carboniferous era, ferns, mosses, and still humbler plants, occupied the throne of the vegetable kingdom, and, by their countless numbers, their huge dimensions, and rank luxuriance, covered the whole earth with a closely-woven mantle of dark green verdure—from Melville Island in the extreme north to the islands of the Antarctic Ocean in the extreme south. The relics of these immense primeval forests, reduced to a carbonaceous or bituminous condition by the secret resources of nature's laboratory, are now buried deep

in the bowels of the earth, packed into solid sandstone cases, and stored up in the smallest compass
by the mighty pressure of ponderous rock-presses,
constituting the chief source of our domestic comfort, and of nearly all our commercial greatness.
A coal-bed is, in fact, a *hortus siccus* of extinct
cryptogamic vegetation, bringing before the imagination a vista of the ancient world, with which no arrangement of landscape or combination of scenery
can now be compared ; and gazing upon its dusky
contents, our minds are baffled in aiming to com-
prehend the bulk of original material, the seasons
of successive growth, and the immeasurable ages
which passed while decay, and maceration, and
chemical changes prepared the fallen vegetation
for fuel.   If the specimens of plants thus strangely
preserved teach us one truth more than another,
it is this, that size and development are terms of
no meaning when applied to a low or a high type
of organization.   The cryptogamia of the old
world, the earliest planting in the new-formed soil,
are in bulk, as well as in elegance and beauty of
form, unrivalled by the finest specimens of the
modern forest.   The little and the great, the recent and the extinct, were equally the objects of
Nature's care, and were all modelled with a skill
and finish that left nothing to be added.

And as in early geological epochs they occupied

so conspicuous a position, so now in the annals of physical geography they are entitled to a prominent place. With the exception of the grasses— Nature's special favourites—they are the most abundant of all plants, possessing inconceivable myriads of individual representatives in every part of the globe, from which unfavourable conditions exclude all other vegetation. And thus they contribute, far more than we are apt from a superficial observation to imagine, to the picturesque and romantic appearances exhibited by scenery, and to the formation of that richly woven and beautifully decorated robe of vegetation which conceals the ghastly skeleton of the earth, and hides from our view the rugged outlines and primitive features of nature. They are the first objects that clothe the naked rocks which rise above the surface of the ocean ; and they are the last traces of vegetation which disappear under degrees of heat and cold fatal to all life. Their structure is so singularly varied and plastic, that they are adapted to every possible situation. In every country they form an important element in the number of plants, the proportion to flowering plants decreasing from and increasing towards the poles. Taking them as a whole, and in regard to their size, they occupy a larger area of the earth's surface than any other kind of vegetation. There are immense forests of

trees here and there in different countries, realizing Cowper's wish for 'a boundless contiguity of
shade ;' there are vast colonies of flowering plants ;
but the range of the most ubiquitous tree or flower
is vastly inferior to that of some of the humblest
lichens and mosses.   Although these plants occupy
but a very subsidiary and unimportant position
among the vegetation which surrounds us in our
daily walks, and are concealed in isolated patches
in the woods and fields by the luxuriance of higher
and more conspicuous plants, yet they constitute
the sole vegetation of very extensive regions of the
earth's surface.   Every part of the globe, within
a thousand feet of the line of perpetual snow, is
redeemed from utter desolation by these plants
alone.   Above the valleys and the lower slopes
which form the step of transition from plain to
mountain—inhabited by prosperous and civilized
nations—is the domain of mist and mystery, the
region of storm—a world which is not of this world,
where God and nature are all in all, and man is
nothing ; and in this unknown region there are immense tracts familiar to the eye of wild bird, to the
summer cloud, the stars and meteors of the night
—strange to human faces and the sound of human
voices, where the lichen and the moss alone luxuriate and carpet the sterile ground.   The grandest
and sublimest regions of the earth are adorned

with garlands of the minutest and humblest plants ;
they are the tapestry, the highly-wrought carpeting laid down in the vestibules of nature's palaces.
The vast mountain-systems of the globe, with
their culminating regions in the Andes, Alps, and
Himalayas, and their subsidiary branches or ribs
in the Grampian, Dovrefjeld, Ural, and Atlantic
ranges, are clothed on their sides, summits, and
elevated plateaus, almost exclusively with cryptogamic vegetation, and enable us to form some con-
ception of the immense altitudinal range of these
plants. Then there are whole islands in the Arctic
and Antarctic Oceans whose vegetation also is
almost entirely cellular. The northern portion of
Lapland, the continent of Greenland, the large
islands of Spitzbergen, Nova Zembla, and Iceland,
the extensive territories of the Hudson's Bay Company, the enormous tracts of level land which
border the Polar Ocean from the North Cape to
Behring's Straits, across the north of Europe and
Asia, and from Behring's Straits to Greenland,
across the north of America, a stretch of many
thousands of miles ; all these immense areas of the
earth's surface—where not a tree, nor a shrub, nor
a flower is seen, except the creeping arctic willow
and birch, and the stunted moss-like saxifrage and
scurvy grass—are covered with fields of lichens and
mosses, far exceeding anything that can be com-

pared in that respect amongst phanerogamous plants. Thus, to the rugged magnificence of Alpine scenery, and the dreary isolation and uniformity of the Arctic steppes, and the boundless wastes of brown desert and misty moorland—to those great outlets from civilisation and the tameness of ordinary life, these humble plants form the sole embellishments.

So much for the distribution of these plants on the land ; their range in the waters is still more extensive. Lichens and mosses cover the waste surfaces of the earth ; diatoms and confervæ are everywhere miraculously abundant in the waters. In rivers and streams, in ditches and ponds, alike under the sunny skies of the south, and in the frozen regions of the north ; on the surface of the sea in floating meadows, and in the dark and dismal recesses of the ocean only to be explored by the long line of the sounding-lead. The ocean swarms with innumerable varieties, without their presence being indicated by any discoloration of the fluid. The Arctic and Antarctic Oceans, covering areas larger than the continents of Europe and Asia, are peopled by myriads of diatoms ; various inland seas and lakes are tinged of different hues by their predominance in the waters ; while it has been ascertained, from the soundings obtained during the investigations connected with

laying the electric telegraph cable between Ireland
and Newfoundland, that the floor of the Atlantic
is paved many feet deep with their silicious shields,
preserving in all their integrity their wonderful
shapes, notwithstanding their extreme delicacy and
minuteness, and the enormous pressure of the vast
body of water which rests above them. Such is
the wide space which these organisms occupy in
the fields of nature—a prominence which is surely
sufficient to redeem them from the charge of insignificance. They are inferior in majesty of form
to palms and oaks, but in their united influence it
is not too extravagant to say that they are not less
important than the great forests of the world.

This vast profusion of minute and humble vegetable life serves the obvious purpose of preparing
the way for higher orders of vegetation. Nature
is incessantly working out vast ends by humble
and scarcely recognisable means. The features
of the earth are being continually altered by the
germination and dispersion of the algæ, mosses,
and lichens. Bare and sterile mountains are
clothed with verdure; rocks are mouldering into
soil ; seas are filling up ; rivers and streams are
continually shifting their outlines ; and lakes are
converted into fertile meadows and the sites of
luxuriant forests, by means of the vast armies of
Nature's pioneers. Hard inorganic matters are re-

duced to impalpable atoms ; waters and gases are
decomposed and moulded into new forms and substances having new properties, by vegetable growth.
Minute as these plants are, they are intimately
related to the giant forms of the universe. It has
been observed that as the great whole is indissolu-
bly connected with its minutest parts, so the germination of the minutest lichen, and the growth
of the simplest moss, is directly linked with the
grandest astronomical phenomena ; nor could the
smallest fungus or conferva be annihilated without
destroying the equilibrium of the universe. It is
with organic nature as with the body politic or the
microcosm of the human frame, " if one member
suffer all the members suffer with it," and the loss
of one class or order would involve that of another,
till all would perish. Our comfort and health, nay
our very existence, more or less immediately depend on the useful functions which these plants
perform. Before we can have the wheat which
forms our daily bread, or the grass which yields us,
through the instrumentality of our herds, our daily
supply of animal food, or the cotton and linen which
form our clothes, countless generations of lichens
and mosses must have been at work preparing a soil
for the growth of the plants which produce these
useful materials. And as on the dry land, so in
the great waters, this wonderful chain of connexion

exists in all its complexity. Before the reader can peruse these pages by the light of the midnight lamp, or the gay party can indulge their revels under the brilliant glare of spermaceti tapers, myriads of minute diatoms and confervæ, floating in the waters of the sea, must have formed a basis of subsistence for the whales and seals whose oil is employed for these purposes. Man's own structure is nourished and built up by the particles which these active plants have rescued from the mineral kingdom, and which once circulated through their simple cells : and thus the highest and most complex creature, by a vital sympathy and a close physical relation, is connected with the lowest and simplest organism, to teach him humility, and inspire him with a deep interest in all the works of his Maker!

It may be asked by a class of individuals, unfortunately too numerous, What is the use of these minute plants to us ? In the business language of the world things are called useful when they promote the profit, convenience, or comfort of everyday life ; and useless when they do not promote, or when they hinder any of these desired ends. But this definition is extremely one-sided. There are higher purposes to serve in this world than mere subservience to the physical wants of man. There is a much higher utility than the mere

mechanical and mercantile one. The useful things
of external life, indeed, should not be undervalued ;
they are the first things required, but they are not
the sole or the highest things necessary. Man
must have food and clothing in order to live ; but
it must also be remembered that man does not
live by bread and the conveniences of external life
alone. When any one does live by these alone, he
has forfeited his claim to the higher form of life
which is his glorious privilege, and by which he is
distinguished from the lower animals. Nature
throughout her whole wide domains gives no countenance to such a materialistic exclusiveness. She
is at once utilitarian and transcendental. Uses and
beauties intermingle. All that is useful is around
us ; but how much more is there beside ? There
is a strange superfluous glory in the summer air ;
there is marvellous beauty in the forms and hues
of flowers ; there is an enchanting sweetness in the
song of birds and the murmur of waters ; there
is a divine grandeur and loveliness in the landscapes of earth and the scenery of the heavens, the
changes of the seasons, the dissolving splendours
of morning, noon, sunset, and night, utterly in-
comprehensible upon the theory of nature's exclusive utilitarianism. All things proclaim that
the Divine Architect, while amply providing for
the physical wants of His creatures, has not for-

gotten their spiritual necessities and enjoyments; and having implanted in the human soul a yearning for the beautiful, has surrounded us with a thousand objects by whose charms that yearning may be gratified. And one of the most striking examples of this Divine care is to be seen in the profusion of minute objects spread around us, which apparently have no direct influence at all upon man's physical nature, and have no connexion with his corporeal necessities. These objects, subserving no gross utilitarian purpose, are intended to educate man's spiritual faculties by the beauties of form, the wonders of structure, and the adaptations of economy which they display. Their beauty is sufficient reason for their existence were there no other. When their varied and exquisitely symmetrical forms are presented to the eye under the microscope, a thrill of pleasure is experienced, calm and pure, because free from all taint of passion, and felt all the more intensely because nameless and indefinite. We are brought face to face with perfection in its most wonderful aspect—the perfection of minuteness and detail; with objects which bear most deeply impressed upon them the signet-mark of their Maker; and we observe with speechless admiration that the Divine attention is acuminated and His skill concentrated on these vital atoms; the last visible organism vanishing from our view with the same Divine glory upon it, as the last star that glimmers out of sight on the remotest verge of space.

These organisms further justify their existence to the utilitarian, inasmuch as their study is well calculated to exercise an educational influence which should not be overlooked or despised. While they try the patience they exercise the faculties by forcing attention upon details. Their minuteness, their general resemblance to each other, their want in many cases of very prominent or marked characteristics, render it a somewhat difficult task to identify them. Long hours may often be spent in ascertaining the name of a single species, and assigning it its proper place in the tribe to which it belongs. One species may often be confounded with another closely allied, and days and weeks may elapse before the eye and the mind familiarized with their respective details, can observe the distinctions between them. This difficulty of identification greatly sharpens one's knowledge, induces a habit of paying attention to minutiæ, and creates a power of distinguishing between things that differ slightly, which is exceedingly valuable and important. For the eye and mind thus educated to detect resemblances and differences in objects, which to ordinary observation appear widely dissimilar or precisely the same, there will be abundant scope in the practical details of common everyday life, as well as in the higher walks of literature, science, and art.

The study of these plants has also a tendency to elevate and enlarge our conceptions of nature ; its vastness and complexity, its incommunicable grandeur, its all but infinity, opening before us newer and more striking vistas with every descending step we take. The further we advance, and the wider our sphere of observation extends, wonder follows on wonder, till our faculties become bewildered, and our intellect falls back on itself in utter hopelessness of arriving at the end. Minute as the objects are in themselves, contact with them cannot fail to excite the mind, to call it forth into full and vigorous exercise, to enlist its sympathies, and to expand its faculties. Many eloquent pages have been written to show this elevating influence upon the mind, of contact with, and contemplation of the phenomena of Nature ; but it is not the great and sublime objects of Nature alone that produce this effect—the sublimity of mountains, the majesty of rivers, and the repose of forests,—the very humblest and simplest objects are calculated to awaken these emotions in a yet higher and purer form. " The microscope," as Mr. Lewes has well observed, "is not the mere extension of a faculty ; it is a new sense."

There are also peculiar pleasures connected with the study of these objects. There is first the pleasure of novelty and discovery—of exploring a realm where everything is comparatively new, and every step is delightful ; where the forms are unfamiliar, and the modes of life hitherto unimagined. There is next the more subtle and refined pleasure of observing the strange truths which they unfold, the beautiful laws which they reveal, and the resemblances and relations which they display. The false romanticism of vulgar fancy requires something pretentious and unnatural to gratify its taste ; but to the true poetical mind, the humblest moss on the wall, or the green slime that creams on the wayside pool, will suggest trains of pleasing and profitable reflection. He who has an observing eye and an appreciating mind for these minute wonders of Nature, need never be alone. Every nook and corner of the earth, howeve barren and dreary to superficial minds, has companions for him ; and on every path he will find what the Indians call a rustawallah, a delightful road-fellow.

To the cryptogamic botanist Nature reveals herself in her wildest, and also in her fairest

aspects. He enters into her guarded retreats—
retiring spots of luxuriant, refreshing, and enticing
beauty, that are hidden from every other eye ;
where the great world of strife and toil speaks not,
and its cares and sorrows are forgotten, and Nature
wakes up the dead divinity within, and rouses the
soul to purer and nobler purposes. The peculiar
haunts of the objects of his search are found on
the sides and summits of lofty mountains, amid
the dark lonely recesses of forests, in the bright
bosom of rivers and lakes and waterfalls, on far-
off unvisited moors, where heaven's serene and
passionless blue is the only thing of beauty, and
in the mossy retreats of dell and dingle, where
Titania and her fays might sport away the dreamy
noontide hours. There he finds the pictures which
the soul treasures most lovingly ; and in these
byways does he gain the truest insight into the
mysteries of life. In thus penetrating into the
very heart of Nature, with much toil and exertion
it may be, he seems to win her confidence, and to
earn the right to look into her arena. By minute
contact and continued commune with her alone
in the wilderness, he feels in all its fulness and
depth the beautiful relationship that exists between
the outer and the inner life of creation. To others
the landscape may be but the mere background of a
picture, in the foreground of which human figures

are acting; to him its charms are agencies and
influences acting on his heart and mingling with
his life. The sportsman in search of game
frequently wanders into regions that seem primeval
in their solitude, and where "human foot has
ne'er or rarely been;" but so absorbing is the pursuit in which he is engaged, that he seldom pauses
to watch the features of the surrounding scenery,
or to notice combinations of objects and effects
of light and shade which nature never displays
except in such unfrequented spots. But to the
cryptogamist, on the other hand, these very scenes
of Nature lend a nameless charm and interest to
the lowly plants he gathers, and are ever after
indelibly associated with them in his memory, and
are renewed every time he witnesses their faded
remains. Hardly a moment passes over the
solitary collector amid such secluded scenes,
without some grand effect being produced in the
surrounding landscape, or in the appearance of
the sky above him; some wonderful transformation of Nature, as though the spot where he stands
were her tiring-room, and she were trying on robe
after robe to see which became her best; some
striking incident, which might well inspire him
with the wish to catch the happy moment, and
give it a permanent existence. Such are the
simple, refining, and enduring pleasures which the

cryptogamic botanist enjoys in the pursuit of his favourite study amid the scenes of Nature.

Add to all these recommendations this last important advantage, that these plants can be observed and collected without interruption throughout the whole year, and in situations where other vegetation is reduced to zero. They can be studied alike under the cloudy skies of December, as when illumined by the sunshine of June. When the flowers and ferns have vanished, when the lights are fled, and the garlands are dead, the deserted banquet-hall of Flora is still relieved by the presence of these humble retainers, whose fidelity is proof against every change of circumstance, and whose better qualities are displayed when the storm is wildest and the desolation most complete. They are no summer friends. As Ruskin has beautifully observed, " Unfading as motionless, the worm frets them not, and the autumn wastes not. Strong in lowliness, they neither blanch in heat, nor pine in frost. To them, slow-fingered, constant-hearted, is intrusted the weaving of the dark eternal tapestries of the hills ; to them, slow-pencilled, iris-dyed, the tender framing of their endless imagery. Sharing the stillness of the unimpassioned rock, they share also its endurance ; and while the winds of departing spring scatter the white hawthorn blossoms like drifted snow, and summer duns in the parched meadow the drooping of its cowslip-gold, far above among the mountains, the silver lichen-spots rest, starlike on the stone, and the gathering orange-stain upon the edge of yonder western peak reflects the sunsets of a thousand years."

CHAPTER I.

MOSSES.

VERY thoughtful mind must be struck with astonishment at the boundless prodigality with which the riches of Nature are thrown broadcast over the whole surface of the earth. The loveliest objects are, as it were, carelessly scattered here and there in waste spots and lonely unvisited haunts, where there is no hand to gather, and no eye to admire them. The great temple of Nature is like the magnificent old temple of Solomon,—upon the top of every pillar is lily-work. The massive and rugged foundation stones of the earth are almost concealed by a profusion of graceful and beautiful things,—the grass, the flowers, the forests ; while the craggy pillars have their capitals enwreathed with exquisite garlands of ferns and mosses. Not a rock peeps above the surface of the soil but has its steep sides clothed with rainbow-hued lichens, and its summit enveloped in verdure. In the smallest and most insignificant of these objects, there is as much of beauty of form and colour and ingenuity of structure displayed as though it were the only object in the universe. Nay, God seems to bestow more abundant honour upon those objects whose smallness and insignificance would otherwise cause them to be overlooked.

Of all the minute flowerless plants with which Nature, as it were, points her flowery sentences, fills up her vacant spaces, and balances and tones her landscapes, mosses are by far the loveliest and the most interesting. As regards form and structure they are the most beautiful of all plants; Nature having bestowed upon them this compensation for want of the varied and gorgeous colouring imparted to the higher tribes of vegetation. Their beauty is not of a glaring or obvious character, but refined and spiritual, consisting in delicacy of tint, in the imperceptible gradation with which one hue is blended with another, in the filmy transparency of the structure, and in the endless diversity and perfection of the form. It is invisible to the careless or the casual observer, but brightens like a star upon the view when attentively and minutely examined, finding an unconscious interpreter in every heart, and affording, when fully perceived, to every thoughtful mind, a purer and more subtle joy than is communicated even by the rose or the lily. Regarded *en masse*, what can be lovelier than a closely-shaven mossy lawn, over which the golden sun-beams, and the light-footed shadows of the fleecy clouds overhead, chase each other throughout the whole summer day in little rippling waves, like smiles and thoughts over a human face! What can be pleasanter than the soft yielding carpets of greenest verdure and weirdest patterns, woven by these tiny plants on the floor of shadowy old forests, "stealing all noises from the foot," and imbuing the mind with reverence and awe in the pillared aisles of Nature's cathedrals! What can be more picturesque than the varied hues which mosses impart to the ivied ruin, the grey old wall, or the decaying tree; or what object can be more romantic than a fantastic rock crowned with pines or birches, with mosses hanging down in waving clusters from its edge, and forming beautiful festoons like draperies of green and brown silk over the pillars of some oriental palace! Truly these little plants originated in a high ideal of creative wisdom and love.

Mosses belong to the foliaceous or highest division of flowerless plants. Although consisting entirely of cellular tissue, and increasing by simple

additions of matter to the growing point or the
apex of parts already formed, they point to far
higher orders of vegetation ; they are prefigura-
tions of the flowering plants, epitomes of archetypes in trees and flowers. There is nothing in
the appearance or structure of the lichens, fungi,
or algæ, to remind the popular mind of higher
plants ; they form, as it were, a strange microcosm of their own—a perfectly distinct and peculiar order of vegetable existence. But when we
ascend a step higher and come to the mosses, we
find for the first time the rudimental characters
and distinctions of root, stem, branches, and leaves
—we recognise an ideal exemplar of the flowering plants, all whose parts and organs are, as it
were, sketched out, in anticipation, in these simple
and tiny organisms. Through the small densely-
cushioned, moss-like alpine flowers, they approximate analogically to the phanerogamous plants in
their leaves and habit of growth ; and through the
cone-like spikes of the club-mosses, they approximate to the pine tribe in their fructification.
From both these classes of highly organized
plants, however, they are separated by wide and
numerous intervening links. But still it is curious
and interesting to find in them an exemplification
of the universal teleology of nature—the humblest
typical forms pointing to the grand archetypes,

the simplest structures anticipating and prefigur-
ing the most highly organized and complicated.

In no tribe of plants is there so great a
similarity between the different species as in the
mosses. In them is strikingly displayed the
grand characteristic feature of God's work in
creation—unity of type with variety of development. A simplicity and uniformity of structure
runs throughout the entire family. The whole
appearance, the general air, the manner of growth,
is the same in all the species ; so much so, that it
is perhaps easier to distinguish a species of moss
than a species of any other plant. This remarkable similarity, concealing a no less remarkable
diversity, has led to the popular belief that there
is only one kind of moss ;—all the species, of
which upwards of five hundred exist in this
country alone, being confounded in one general
appearance. Closely examined, however, by an
educated eye, their exceeding variableness of
form will at once become evident, some being
slender hair-like plants ; some resembling miniature fir-trees, others cedars, and others crested
feathers and ostrich plumes. In size they vary
from a minute film of green scarcely visible to the
naked eye, to wreathes and clusters several feet
in length. Nor are their colours less variable,
ranging from white, through every shade of

yellow, red, green, and brown, to the deepest and
most sombre black.

Though most of the peculiarities of mosses are
visible to the naked eye, it is on the stage of the
microscope that they appear to the greatest advantage. The modifications of structure to suit
the requirements of their economy thus revealed,
cannot fail to excite our admiration and astonishment. The stems of mosses, though serving the
same purposes, are widely different from those of
flowering plants. We are ignorant of the manner
in which they are developed. Probably, like en-
dogenous plants, which is the least complicated of
the two natural processes of increase in the vegetable kingdom, they grow by successive additions
to the summit, never increasing the diameter after
their outer layer has been formed. They are solid,
and composed entirely of cellular tissue, which
gradually becomes softer and more porous near
the centre, uniform in every part, having neither
medullary rays, nor true outward bark, nor central
pith, nor even the scalariform vessels observable
in the stems of ferns. Of the course taken by the
ascending and descending sap, we are equally
ignorant, if indeed there really exist in them
currents similar to those of flowering plants, which
may be more than doubted. The roots are exceedingly delicate organs, and yet they take as

firm a hold of the earth, in proportion to their size, as the roots of many trees. In some cases they consist of small thread-like fibres, or long creeping underground stems ; while in others they are aërial, like those of orchids, being developed in the form of a thick silky down of a pale brown colour, imbedded among the leaves close to the stem. This last variety of root is to be seen chiefly in species that grow in moist or watery places, where they act as sponges to attract and preserve the humidity of the plants, when the moisture around them is dried up. In connexion with their roots we observe a striking provision of Nature for the welfare of mosses in unfavourable circumstances. As the most delicate fibres hardly penetrate beyond the surface of the soil, which in dry sultry weather speedily parts with its moisture, the mosses would perish were they entirely dependent for their nourishment upon their roots. But every part of them, and especially the leaves, is endowed, to a remarkable degree, with the power of imbibing the faintest moisture from the air, and reviving, even when apparently withered and dead, on the recurrence of a shower of rain. The roots therefore, in most instances, serve only to attach the plant to its growing-place, the functions of nutrition being performed indiscriminately by its whole surface.

The leaves of mosses are their most prominent parts. To the careless and superficial eye, accustomed to look at a tuft of moss as merely a patch of velvety greenness, creeping over an old tree or dike, the leaves of all mosses may appear precisely similar ; but the attentive observer who examines them under a microscope, will find that the leaves of different kinds of trees are not more distinct from each other than are those of the mosses. Indeed, so remarkable and so constant is this dissimilarity, that it has formed one of the principal bases of their arrangement and classification ; and the botanist who has studied them thoroughly can identify under the microscope, in some cases, the smallest fragment of a leaf, although almost invisible to the naked eye. The leaves of some mosses are quite plain and pellucid, exhibiting no structural arrangement whatever ; others are furnished with a nerve which runs through the centre and terminates above or below the apex ; some are either ribbed and notched like a saw on the edge, or quite plain and even ; and others present the most beautiful and varied network of cells. Some are linear like miniature pine-needles, others ovate and round like the leaves of our common deciduous trees. The harmonies of colours are beautifully exhibited in their appendicular parts. The stem, in almost all the species, is of a pale wine-red colour, while the leaves are generally of a delicate pea-green hue. In some species the leaves are of the deepest and most vivid green, while their margins and nerves are of a deep blood-red colour. The fruit-stalk and fruit-vessel are sometimes red or orange-coloured, while the leaves are brown ; and sometimes dark brown, when the leaves are of a golden yellow. Unlike the leaves of ferns, which are mere foliaceous expansions of the stem, and developed in one plane, the leaves of mosses are quite distinct from the stem, and are arranged around it on all sides, most frequently in an alternate manner, so that a line joining their bases would form a spiral more or less elongated.

The organs of fructification, however, with which mosses are furnished, are perhaps the most wonderful parts of their economy. When the requisite conditions are present, these are generally developed during the winter and spring months, and may be easily recognised by their peculiar appearance. At first a forest of hair-like stalks, of a pale pink colour, rises above the general level of the tuft of moss, to the height of between one and three inches, giving to the moss the appearance of a pincushion well provided with pins. These stalks, through course of time, are crowned with little urn-like vessels called capsules, which are covered

C

at an early stage with little caps, like those of the
Normandy peasants, with high peaks and long
lappets,—in one species bearing a remarkable
resemblance to the extinguisher of a candle,—a
curious provision for protecting them alike from the
sunshine and the rain, until the delicate structures
underneath are matured. When the fruit-stalk
lengthens and the capsules swell, this hood or cap
is torn from its support, and carried up on the top
of the seed-vessel, much in the same way as the
calyx of the common garden annual, the *Esch-
scholtzia or Californian Poppy* is borne up on the
summit of the cone-like petals before they expand.
When the seed-vessel is riper it falls off altogether,
and discloses a little lid covering the mouth of the
capsule, which is also removed at a more advanced
stage of growth. The mouth of the seed-vessel is
then seen to be fringed all round with a single or
double row of teeth, which closely fit into each
other, and completely close up the aperture. It
is a circumstance worthy of being noticed, that the
even numbers which prevail in the formation of
microscopic cells, are also found in these organs,
the teeth being arranged in each row in the geo-
metrical progression of 4, 8, 16, 32, or 64, there
never being by any chance an odd number; thus
illustrating the general doctrine that a system of
types runs throughout the whole works of Nature,

furnishing evidences of supreme intelligence, and wonderfully adapted not only to the objects to which it is applied, but also to the same or similar principles in the constitution of man's mind.

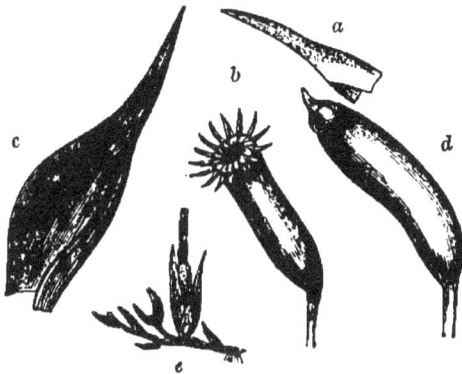

FIG. 1.—Bryum serpens.
(a) Veil. (b) Fringe. (c) Leaf. (d) Capsule with lid. (e) Stem.

These teeth are highly sensitive to the changes of the weather, opening in sunshine, and closing during moist or rainy weather, for the obvious purpose of ripening the minute dust-like seeds with which the interior of the capsule is filled ; and it is a remarkable circumstance that, in one or two genera of mosses which are not provided with hygrometric teeth, the lid that closes the capsule is permanent, being thrown off only when the seeds are ripe and ready to be dispersed. By placing a capsule, the teeth of which are closed, near the fire or in the warm sunshine, the teeth will be seen to open with a graceful and gradual

motion ; while the slightest moisture of one's
breath invariably causes the little teeth instantly
to close over the mouth.   This extremely simple
mechanism is one of the most wonderful contri-
vances of Nature, one of the most extraordinary
adaptations of means to an end, to be found in
the whole economy of vegetation.   Within the
capsule the seeds surround a slender pillar or columella, and are enclosed in a membranous bag.
The seta or fruit-stalk is in some mosses terminal,
and in others lateral, springing from the top of the
stem or the side, and these characters afford a
convenient mode of arranging the whole tribe into
acrocarpi or *pleurocarpi.*   Mosses with terminal
fruit-stalks are more fugacious, and may be regarded as analogous to spring or annual plants
whose blossoms are produced at the summit of
the stem or direct from the root ; while mosses
with lateral fruit-stalks are more permanent, and
may be compared to perennial plants, shrubs,
or trees, whose blossoms or fruit appear on side
sprouts or branches, and are concealed among
the foliage.   Elevated as the seed-vessels are by
their stalks, they are freely exposed to the ripening effects of sun and wind ; and it is a curious
sight to see these straight footstalks gradually
bending, reversing the seed-vessels, and empty-
ing the seeds they contain as from a pitcher,

to be carried by the wind to some congenial spot, where through course of time they may spring up and form a new colony of mosses, which in their-turn will carry on the circle of life, from the seed to the full-grown moss, and from the full-grown moss to the seed, the beginning and the ending, the ending and the beginning.

Besides these curious capsules, there are other organs of fructification which clearly demonstrate the sexuality of mosses. Their real nature has only recently been accurately ascertained. They are called antheridia and pistillidia or archegonia, from the strong resemblance which they bear to the stamens and pistils of the flowering plants, and from their being supposed to perform the same or analogous functions. They are small spherical or flask-shaped bodies, fixed by short footstalks, concealed in cup-shaped receptacles among the perichætial or uppermost leaves, and often occur in abundance along with the capsules on the same plant. Examined under the microscope, the antheridia are found to consist of a bag, whose membrane is formed of somewhat oblique cells, containing granular matter arranged around a bright red nuclear body, which divides into a number of small vesicular bodies of precisely the same character. This granular matter, under a higher power of the microscope, is resolved into a

mass of apparently living animalcules called spermatozoids or antherozoids. These tiny organisms
have short slender bodies, with long spirally-twisted
tails, and display the most lively movements, each
whirling upon its own axis, and quickly running
about the field as if from an intense feeling of
sensuous enjoyment. These movements generally
cease in the course of two hours after the discharge
of the spermatozoids from the antheridia ; but
sometimes they are observed to move actively
even after the lapse of two days. They are furnished with cilia like animalcules ; and their motion
is such as would undoubtedly be attributed to
ciliary action if seen in an animal structure. They
are nothing more, however, than mere modifications
of vegetable tissue ; and their motion is simply a
hygrometrical action, like that of the teeth which
fringe the mouth of the capsule. They are extremely curious objects, and well worthy of the
most careful examination. In the same receptacle
among the upper leaves of the moss, may be seen
antheridia in every stage of development, those in
the centre appearing to ripen first, even while some
of those at the outer edge are of small size and
quite green. There is thus a constant succession
of spermatozoids produced ; a provision which
tends to insure their application to the pistillidia
at the proper time.

Several species of mosses are furnished with pseudopodia, which consist of powdery or granulated heads terminating an elongated and almost leafless portion of the stem. These organs are usually developed only in unfavourable circumstances, being formed at the expense of the fruit, which is then abortive. They appear to be simply a mass of naked seed, without the ordinary protection and mechanism of an enveloping seed-vessel, and as such, afford a remarkable illustration of the simplicity of the means by which nature, when placed at a disadvantage, effects her vital purposes. Several mosses, however, possess the power of maintaining and spreading themselves without the aid of any of these organs of fructification ; thus showing that the conditions essential to the act of reproduction in the higher ranks of creation may be gradually dispensed with as we descend the scale, until they are at length altogether superseded by the simplest process when we reach the extreme limits either of the animal or vegetable kingdom. There is one remarkable moss, the male plant of which exists only in Europe, so far as can be ascertained, and the female only in America, and yet they propagate themselves by a process of proliferous growth or budding with as much facility as though they grew side by side in the same crevice of rock. Almost all the mosses,

which cover extensive areas of mountain and lawn,
and occupy large tracts of bogs and watery wastes,
are barren; it being a rare thing to find on them
capsules or any of the other compensating organs.
They are exceedingly proliferous, throwing out
young shoots from their sides or summits, and thus
often increasing many feet in depth, forming layer
above layer, the uppermost stratum alone being
vital; the rest decomposed into peat, forming a
rich organic soil for its nourishment. This process
of multiplication among the mosses is analogous to
the process of budding in the higher plants by which,
without sexual elements, without blossom or seed,
a perennial plant, such as a tree, produces new individuals from the development of its leaf-buds.

It is extremely interesting to notice that the
leaf is the type of the plant in the moss as in the
flowering plant; the veil being merely a convolute
leaf, the lid a metamorphosed leaf, the teeth one
or more whorls of minute flat leaves. It is by no
means rare to find individual mosses in which
leaves appear at the top of the fruit-stalk in place
of the spore-case, just as happens in the phyllody
of flowering plants, when the coloured parts of the
flower are converted into green foliage.

Mosses possess in a high degree the power of
reproducing such parts of their tissue as have been
injured or removed. They may be trodden under

foot ; they may be torn up by the plough or the harrow ; they may be cropped down to the earth, when mixed with grass, by graminivorous animals ; they may be injured in a hundred other ways ; but, in a marvellously short space of time, they spring up as verdant in their appearance and as perfect in their form as though they had never been disturbed. The necessity of such a power of regeneration as this is abundantly manifest, when we consider the numberless casualties to which they are exposed in the bare shelterless positions which they occupy.

Mosses also possess the power of resisting, perhaps to a greater extent than most plants, the injurious operation of physical agents ; and this likewise is a wise provision to qualify them for the uses which they serve in the economy of nature. The influence of heat and cold upon many of them is extremely limited ; some species flourishing indiscriminately on the mountains of Greenland and the plains of Africa. They have been found growing near hot springs in Cochin-China, and fringing the sides of the geysers of Iceland, where they must have vegetated in a heat equal to 186 degrees ; while, on the other hand, they have been gathered in Melville Island at 35 degrees, or only just above the freezing-point. Though frozen hard under the snow-wreaths of winter for several

months, their vitality is unimpaired ; and though
subjected to the scorching rays of the summer's
sun, they continue green and unblighted. Even
when thoroughly desiccated into a brown un-
shapen mass that almost crumbles into dust when
touched by the hand, they revive under the influence of the genial shower, become green as an
emerald ; every pellucid leaf serving as a tiny
mirror on which to catch the stray sunbeams.
Specimens dried and pressed in the herbarium for
half a century, have been resuscitated on the application of moisture, and the seed procured from
their capsules has readily germinated. They grow
freely in the Arctic regions, where there is a long
twilight of six months' duration ; and they luxuriate in the dazzling uninterrupted light of the
tropics. They are found thriving amid moist
steam-like vapours, with orchids and tillandsias,
in the deep American forests ; and they may be
seen in tufts here and there on the dry and arid
sands of the Arabian deserts. It matters not to
the healthy exercise of their functions whether the
surrounding air be stagnant or in motion, for we
find them on the mountain top amid howling
winds and driving storms, and in the calm, silent,
secluded wood, where hardly a breeze penetrates
to ruffle their leaves.

The range of flowering plants is circumscribed

by conditions of light, temperature, elevation
above the sea, geological character of the district,
and various other physical causes ; but the wonderful vital energy with which the mosses are endowed, enables them to resist the most unfavourable influences, to grow freely and luxuriantly
even in the bleakest circumstances, and to accli-
matize themselves, without changing their character, in any region of the earth, and every kind of
situation upon its surface ; while, owing to the
extreme minuteness and profusion of their germs
of reproduction, they are almost universally disseminated by the winds and waves. There is no
spot so barren and desolate where some species
or other may not be found. Although often growing in great abundance within the tropics, carpeting the ground, and covering the trunks of the
trees, and sometimes attaining very luxuriant proportions, the temperate zones, however, are the
proper regions of the mosses. Unlike the ferns,
the size and number of which gradually diminish
in passing from tropical to temperate countries,
the maximum of mosses is found in cold climates,
increasing in luxuriance, beauty, and abundance
as we approach the North Pole. Like the ferns,
moisture and shade are highly favourable to their
growth and wellbeing ; hence, as a rule, they produce a larger number of species and individuals,

and spread over wider areas in islands and the
vicinity of rivers and lakes than in the interior of
continents, unless when well wooded and watered.
Their favourite habitats appear to be rocky dells or
ravines at the foot of mountains, with streamlets
murmuring through them, and dense trees inter-
weaving their foliage over their sides, and creating
a dim twilight in the recesses beneath. In such
hermit seclusions the botanist may expect to reap
the richest harvest of species.

Mosses occasionally select very singular places
of growth ; and notwithstanding the minuteness
and profusion of their seeds, the facility with
which they can be disseminated, and their insensibility to ordinary physical conditions, are, specifically considered, sometimes very much restricted
in their geographical range. Several kinds are
found in this country only on the summits of the
highest Highland mountains, covering the barren
soil with a thin film of verdure, or creeping over
the weather-beaten rocks in tenacious dark-col-
oured clusters or tufts. These species are identical with those found on the plains of the Arctic
regions and the hills of Lapland and Greenland,
where they occur not merely in isolated tufts, as
we find them in this country, but carpeting the
ground for many yards, and imparting a verdant
hue to the mountains and valleys. This circum-

stance would indicate that their original centre of
distribution exists in these dreary regions, and
that from thence they have been disseminated
over the British and European mountains. The
Alpine species are exceedingly restricted, seldom
being found lower than 3000 feet, and often ascending to a height of 4000 feet on the British hills,
and 8000 feet on the Alps of Switzerland and the
Pyrenees ; the isothermal line of these altitudes
corresponding with the plains of Lapland and
the level of the sea-shore in the Arctic regions.
Along with the small moss-like Alpine flowers
with which they grow, they must have been wafted
down to the Highland mountains, either as germs
or as full-sized plants, growing undisturbed in
their native soil, when these mountains existed as
islands in the midst of an immense glacial sea
which swept over what is now the continent of
Europe. When this sea retired, owing to the
elevation of the land, and its islands became
mountain peaks and ranges, the tiny plants which
imparted to them their first faint tinge of verdure
still remained, finding the same conditions of
temperature, shade, and moisture among the
clouds as they formerly found on the shore of an
icy sea. Thus all the Alpine plants found on the
summits of our loftiest hills are Norwegian or
Arctic species. They are besides the oldest living

plants in the world, each of them, even the very
humblest moss or saxifrage, having a pedigree
which extends into the misty past, unknown ages
before the creation of man. What an intense,
almost human interest, gathers around these tiny
mosses and fragile flowers, which bloom like lone
stars in a midnight sky, in the very hoof-marks of
the storm, when we reflect that they are the last
of their race, the scanty remains of what was once
for many ages the general Flora of the whole of
Europe. True patriots, they have clung to their
native homes, although they have changed their
very nature; retiring before the inroads of the
host of gaudy flowers which invaded our valleys
and woods from the East, to the storm-scalped
summits of the Highland mountains, and behind
the icy battlements of the Arctic regions. Upwards of fifty species are confined to the lofty
ranges in the centre of Scotland, especially the
Braemar and Breadalbane mountains, which form
the most important part of the great Grampian
range, and contain the most extensively and
uniformly elevated land in Great Britain. These
species are pre-eminently Arctic and Norwegian,
and present many striking peculiarities which distinguish them at a glance from the mosses of the
woods and the valleys. Though confined to the
shoulders and the summits of our loftiest moun-

tains, they are common hyperborean mosses, growing most luxuriantly and spreading in wide patches on the rocky plains of Spitzbergen, and in the upland woods of northern Norway. A few of them are found on the highest mountains of Wales and the south of Ireland; while the remaining representatives of these Alpine and Arctic mosses cover the projecting rocks which tower up through the glaciers of the Alps and the snows of the Pyrenees. No less than a dozen are exclusively restricted to the very highest summits of the most elevated peaks in Britain, never, except when brought down by streamlets in isolated tufts along their course, descending to a lower altitude than 4000 feet; while upwards of forty of the rarest species are found on Ben Lawers and the lofty hills in the neighbourhood, of which no less than twenty are to be found nowhere else in this country. On Ben Lawers alone 330 species occur.

Some mosses are very much restricted in their range. The *Glyphomitrion Daviesii*—a minute darkish-green moss spreading over rocks generally near the sea—is peculiar, so far as known, to the British Isles, where it is exceedingly local, being found in one or two places in Ireland and Wales, and in Scotland on the trap rocks at Bowling on the Clyde. The lovely curve-stalked apple-moss (*Bartramia arcuata*), which covers moist banks and rocks in alpine districts in this country with its rich golden foliage and thick ferruginous downy roots in the utmost profusion, appears to be wholly unknown upon the Continent. While, on the other hand, the feathered Neckera (*Neckera pennata*), which is not uncommon in Switzerland, has been found in only one station in Great Britain, viz., on the trunk of a beech at Fotheringham, near Forfar, by Mr. Drummond. The same may be said regarding Haller's feather-moss (*Hypnum Halleri*), which is abundant in Switzerland, and in this country occurs only on rocks near the summit of Ben Lawers. Two summers ago I discovered, on the western shore of Loch Corruisk, in the Isle of Skye, great quantities of the *Myurium Hebridarum* of Schimper, a moss which had previously been found only once before in South Uist, one of the Hebrides. It is somewhat abundant in the Azores and in Madeira. In habit and appearance it differs totally from all the British mosses, and resembles many of the New Zealand species. It has a decidedly foreign look, and is exceedingly beautiful, bearing some resemblance, in its crisp, glossy, silken foliage, to the *Neckera crispa*, only that it grows in tufts on the ground, instead of in long pendent wreaths. The fact of such a conspicuous moss having remained so long unknown, although at Corruisk

it carpeted a pathway worn smooth by the feet of
travellers winding along the shores of the lake,
indicates that several species, known to occur in
very few places in Europe, may be expected to
be found on the south or western coasts of the
British Islands. The Myurium belongs to a small
group of ferns, mosses, and lichens, such as the
Trichomanes radicans or transparent bristle-fern
of Killarney, the *Sticta macrophylla* or broad-
leaved lichen, and the *Hookeria læte-virens*, a
species of moss found sparingly in the south-
west of Ireland, and common and abundant in
Madeira and the Azores. Along with the Iberian
flowering plants found in the same Irish locality,
they bear witness to the fabled Atlantis, or
hypothetical continent in the Atlantic, by which
the three Macaronesian groups of Madeira, the
Canaries, and the Azores were connected with
western Europe and north-western Africa. There
is a remarkable coincidence, regarding this hypothetical continent, between the facts of natural
history and the traditions handed down on both
sides of the Atlantic, and embodied in the works
of Plato and Theopompus, and in the Teo Amoxtli
of Mexico, as translated by the Abbé Brasseur
de Bourbourg. One exceedingly singular custom
called the *couvade*, in which the father is put
to bed on the birth of a child, existing in both

hemispheres, amongst the Carib races of America, and the Iberians in the North of Spain, points in the same direction as the Iberian flora of Ireland and the *Myurium* of Madeira and the Hebrides. The more thoroughly we collate the isolated testimonies of natural history, and the traditions of ancient nations and tribes in Europe and America, and the more clearly we understand the phenomena which took place on the melting of the ice at the close of the glacial period, and map out the limits of ancient glacial action all over the world, the more confirmed, I believe, we shall be in the reality of the fabled Atlantis. Here is a noble inquiry, an unexplored vista of grand research, opened up by the suggestions of a minute moss growing beside a Highland lake.

Some species of moss are very widely distributed. The beautiful proliferous Feather-moss (*Hypnum proliferum*) is an inhabitant of nearly every part of the world. It is almost as well known in Tropical lands as in our own pine forests; and while among the commonest, it is at the same time one of the loveliest and most graceful of the whole tribe. Nothing can exceed the delicacy and intricacy of its filigree patterns, and the richness and variety of its shades of green colour. The cypress-leaved Feather-moss (*H. cupressiforme*), which is extremely common and abundant with us on banks and trunks of trees, has a range of distribution equally wide. Being the most sportive of all mosses, .it can adapt itself, with slight modifications of shape and structure, to almost every variety of conditions. In Madeira and Teneriffe it is the most abundant and sportive of mosses, covering the trunks of trees, especially of the laurel trees in the grand old evergreen woods of the central mountain range, with a most luxuriant drapery. A painter's eye would be delighted with the picturesque appearance of the trunks and boughs of the trees in the moist, dark ravines, cushioned thickly with this moss. The *Neckera crispa,* exceedingly abundant in our own subalpine woods, grows in great profusion in the laurel forests of Madeira, especially at Ribeiro Frio, often entirely covering trunks and branches, and clothing whole trees in mossy drapery. *Pterogonium gracile* and *Polytrichum juniperinum* are also as common in all the Atlantic islands as in our own country. Two species of moss, the *Astrodontium Canariense* and *Neckera intermedia,* as far as known to us, are peculiar to the Atlantic islands ; while one species, the *Glyphocarpus Webbii,* has hitherto been found nowhere else than in Teneriffe, where it occurs in great abundance, covering the moist rocks with broad cushions of a rich yellow hue. There are

three species of moss in Madeira which belong
to the Leskeoid group of Hypnum or Feather
mosses, which are very remarkable in this respect
that it is only in Australia, Tasmania, and New
Zealand that two entirely congeneric species are
known to grow. The connexion between these
closely allied species of the same group, separated
by the whole breadth of the globe from each other,
is a puzzling circumstance. : Well-developed
stems, forming tree-like branches of one of these
mosses (*Leskea spinosa*) often exceed six inches
in height. On the Cameroon mountains, at a height
of 8000 feet, among abundance of our own common mosses, *Polytrichum juniperinum* and *Funaria
hygrometrica*, occur two lovely transparent green
mosses, *Hookeria læte-virens* and *H. splachnoides*,
which are also found at the Lakes of Killarney
and in Madeira and the Azores. While in the
south-west of Ireland occurs the *Adelanthus* or
Jungermannia decipiens, a Scale-moss, which has
also been gathered in St. Helena, Fernando Po,
Quito, and Monte Tunguragua in Peru. From
these few examples, and the number might be
greatly increased, it will be seen that mosses are
almost as important as flowering plants in solving
the problems of the geographical distribution of
plants.

Mosses, in many instances, are limited in their

range to rocks and soils of the same mineral
character; their limits of distribution, and of the
rocks and soils possessing such character, being
identical. For instance, some are confined to
limestone districts and chalk cliffs; a calcareous
soil being indispensable to their existence. Others
affect granite; numerous species luxuriate in soil
formed by the disintegration of micaceous schist;
while not a few are found growing chiefly on sandstone and clay. Some are found only on and near
the sea-shore; others are confined to the beds of
streams and cliffs moistened by the spray of cascades, where, however impetuous the torrent may
be, they cling tenaciously to the rocks, and form
carpets of greenest verdure for the white glistening
feet of the descending waters. Some are restricted
exclusively to trees, whose trunks and boughs
they clasp like emerald bracelets; others lead a
lonely, hermit-like existence, in the dim moist
caves and crevices of rocks, where they are discovered only by the glistening of a stray ad-
venturous sunbeam on the drops of dew trembling
upon their shining golden-green leaves. One
species has actually been found covering the half-
decayed hat of a traveller who had perished in a
storm on Mont St. Bernard. There is a very
peculiar genus called *Splachnum*, whose members
are only found on organic remains, on the blanched

and polished skulls and bones of hares and sheep which had furnished a meal to the fox or the eagle, or on droppings of game and cattle which browse upon the higher hills. This is the only vegetable we find to be contemporary with or posterior to the creation of animals, with the exception of minute microscopic entophytes which grow within the bodies of men and the lower animals. There was an obvious necessity for the universal precedence of plants in creation, for the hard inorganic elements of the rocks had first to be converted by the vital energies of plants into organic substances, before animals could be sustained. It is true that the first created plant and the first created animal derived their origin alike from the inorganic soil, and were endowed alike with the power of converting heterogeneous matter into their own proper substance. But here the resemblance between them began and ended. The plant still possesses its original power of deriving its nourishment from the soil, while the animal has no such power, and is dependent for its support upon matter previously organized to a certain degree by the plant. Thus it is the peculiar function of the plant to effect that important change by which inorganic matter is converted into living substance ; it is in the organs of the plant that matter becomes vital. This is by far the most wonderful operation that goes on in the world ; for in all that afterwards takes place there is no such radical change, there is simply development into more highly organized substance. Yet in what the operation consists, or by what process it is accomplished, is involved in the greatest mystery.

Mosses are sometimes found in an isolated state as single individuals, but they are far oftener found in a social condition. It is a peculiarity of the family to grow in tufts or clusters, the appearance of which is always distinct and well-marked in different species, and often affords a specific character. This disposition to grow together, which is exhibited in no other plants so strongly, redeems them from the insignificance of their individual state, and enables them to modify in many places the appearance of the general landscape. As social plants they often cover vast districts of land. Along with lichens they give a verdant appearance to the desert steppes of northern Europe, Asia, and America. Mixed with grass they luxuriate in parks, lawns, and meadows, particularly in moist, low-lying situations. They spread in large patches over the ground in woods and forests ; and at a certain elevation on mountain ranges, they take exclusive possession of the soil, forming immense beds into which the foot sinks up to the ankle at every step, bleached on

the surface by the sunshine and rain, blackened
here and there by dissolving wreaths of snow
which lie upon them through all the summer
months, and gradually decomposing underneath
into black vegetable mould. The shoulders,
ridges, and elevated plateaus of all the Highland
mountains are covered with huge luxuriant masses
of the woolly-fringe moss [1] (*Trichostomium lanugi-
nosum*), growing continuously over whole acres of
ground, and banishing every other plant from its
domains. Mountain peat, which is of a dry, friable
nature, is formed almost exclusively by the decay
of this moss. It seems intended by nature to
serve as a covering to the soil—in the absence of
grass and heather—as it is found most luxuriantly
and in the greatest profusion in spots considerably
above the heather line, and even above the point
where grass ceases to be a social plant, and occurs
only in scattered tufts here and there. In these
bleak and desolate spots, it sometimes furnishes
materials for an extemporaneous couch to the be-
lated traveller, compelled to sleep in the shade of
a rock on the hills ; although care must be taken
in arranging the couch to place the dry surface
uniformly uppermost, otherwise the wet decomposed portions will here and there obtrude, and
render the repose of the tenant exceedingly un-

[1] See Frontispiece.

comfortable. The common hair-moss[1] (*Polytri-chum commune*), which is the strongest and wiriest of the British mosses, often covers large tracts of moorland, in moist places, and frequently attains a height of between two and three feet. In Lapland it forms almost the only verdure of the plains, and is occasionally used by the inhabitants when on long journeys for a bed ; a large portion of the mossy turf, cut from a neighbouring spot, being employed as a covering. The fountain apple-moss (*Bartramia fontana*) also grows in great profusion wherever it occurs. It completely fills up the sources of springs, for many yards around, with a bright green deceptive verdure, through which the unwary foot sinks into the coldest water and the blackest mud. The course of Alpine streamlets, near their commencement, may be traced for a considerable distance by the beds of this moss, through which the waters languidly flow. But of all the members of this family the *Sphagna*[1] or bog-mosses are the most social. They are everywhere most abundant on heaths and mossy soils, where they spread in such immense masses that they give a singularly light appearance to the whole moorland landscape ; and by the accumulation of their remains fill up the beds of ancient lakes, bogs, and marshes, with dense, spongy, con-

[1] See Frontispiece,

tinuous cushions, of a pale green, dirty white, or
dark red colour. This is the principal moss in
the marshy plains of Lapland, and within the
whole of the Arctic Circle ; and nothing can be
more dreary and desolate than the scenery where
this moss exclusively prevails. Melville Island,
one of the most advanced western points navigated
in the Polar Sea, though nearly as large as Scotland, is principally covered with mosses, these
plants forming more than a fourth part of its whole
flora ; while the black lifeless soil of New South
Shetland, one of the most southern points in the
Antarctic regions, is covered with faint specks of
mosses struggling for existence. In the extreme
north and the extreme south, they thus form the
principal vegetation of large portions of the earth's
surface.

Mosses are seldom associated with historical or
personal incidents. There are two species, however, which derive an additional interest from this
connexion. It has been supposed that the hyssop
which formed the lowest limit in the descending
scale of Solomon's botanical knowledge, and which
was frequently employed in the temple service of
the Jews for purposes of purification by water or
blood, is identical with the little beardless moss
(*Gymnostomum truncatulum*), which is abundant
on banks, walls, and fallow fields in this country.

Others have conjectured that the Caper-plant, *Capparis spinosa,* is the species alluded to. But on a subject so much disputed I venture no dogmatic opinion. The moss in question has been found in little scattered tufts on the walls of Jerusalem, the kind of situation indicated in Scripture as the natural growing-place of the hyssop. It is little more than half-an-inch in height, but it is very much branched, and forms sometimes large continuous patches, which could easily be employed as sponges. The specimens found in the East are considerably larger than those which occur in this country ; so that there is a certain verisimilitude in the reference of Hasselquist, who called it *Hyssopus Solomonis.* The moss which so deeply interested the feelings of Mungo Park in the African desert, as to revive his drooping spirits when overcome with fatigue, has been found, by means of original specimens, to be the little fern-like fork-moss (*Fissidens bryoides*),[1] a frequent denizen of moist banks in woods in this country, although, from its very minute size, often overlooked. There is one peculiar species, the cord-moss (*Funaria hygrometrica*), called *la charbonière* in France, from its growing in the woods where anything has been burned, and particularly abundant on old walls, whose stem possesses the curious hygrometric action observable

[1] See Frontispiece.

in the teeth of other species. In dry weather it becomes corded, while it uncoils and straightens in moist weather, and thus forms an excellent natural hygrometer. As particular illustrations of the beauty of mosses, which can be perfectly seen and appreciated by the naked eye, may be instanced the *Splachnum rubrum* of the North American bogs, with its large, bright red, flagon-shaped fruit-vessel, and its broad, pellucid, soft green leaves ; the common long-leaved thyme-moss[1] of our own woods, with its exquisite, prominent undulated foliage, like a palm-tree in miniature ; and the *Neckera crispa*, which is perhaps the loveliest of all the species, investing rocks and trunks of trees with its richly-coloured and glossy leaves. When spreading over trees, it is of a dark, dull green colour ; but when occurring on dry lichen-clad rocks, over which its closely-adhering stems and leaves creep for many a yard, it assumes a bright yellowish green, glossy hue, changing gradually and imperceptibly downwards, until the old leaves become of a singularly rich dark brown or red colour. When the sunbeams and shadows are flickering over its crisped and silken leaves, it forms one of the most beautiful objects upon which the eye can rest.

Several mosses are distinguished for their curious appearance or structure. One of the queerest of

[1] See Frontispiece.

our British mosses, named after a queer old German
who first detected the plant in Russia, is the
Buxbaumia aphylla—first discovered by Sir William
Hooker in England at Sprouston, near Norwich,
in a fir plantation, and occurring generally on the
ground in fir woods throughout Britain, though
very local and nowhere abundant. From its minute
size it is apt to be overlooked, even where it
grows. Its stem is reduced to a little conical
bulb clothed with minute scales, which are rudimentary leaves, from which rises up a red tuberculated
fruit-stalk about an inch high, bearing a large ovate
oblique capsule crowned with a minute mitre-like
veil. When this veil falls off, the mouth of the
capsule is seen to be fringed with three rows of
teeth ; the inner row forming a plaited or twisted
membranous cone like the teeth of the Tortulas ;
the middle row consisting of numerous erect joint-
less teeth ; and the outer assuming the shape of a
persistent annulus or ring, with gracefully recurved
divisions like the petals of a Tiger-lily. When
the spores in the interior of the capsule are ripe, its
walls give way on one side, and falling off, expose
the spore-sac, which looks not unlike the flower of
the *arum* or the *calla* within its spathe. The spore-
sac also bursts to discharge the spores, leaving the
large columella, with the persistent lid of the capsule on its top, on the fruit-stalk. This moss

differs in structure and appearance from all other
mosses. It is as great an anomaly among mosses
as a Rhizogens, such as the Rafflesia or Balano-
phora, is among flowering plants. An attentive
examination of its peculiar structure will amply
reward the microscopic student. Closely allied to
it is another British moss, the *Diphyscium foliosum*,
almost equally curious. It is not uncommon on
banks and old wall-tops in alpine situations. It is
a minute plant, with no stems, hair-like leaves, and
very large oblique pale-yellow capsule nestling
among the leaves. Indeed, so large is the fruit in
proportion to the size of the plant, that it may be
said to be all capsule together. By that peculiarity
alone it may be known from all other mosses.
The leaves are of two kinds; the lower being filled
with chlorophyll and the upper being destitute of
that substance, and therefore looser in texture. The
individual plants are scattered over the turfy bank,
each little tuft producing its own one capsule.

In mosses we have the same gradation in the
scale of development that we observe in the flowering-plants. As Phanerogamous plants advance in
point of organization and form from grass to
deciduous trees, from the humblest wayside weed
to the giant oak of centuries, so mosses rise in type
and size from the minute Phascum or Earth-moss,
which forms a mere film of green upon the ground,

to the highly complex Hair or Feather-moss, which forms a miniature forest in the depths of the wood, and grows in some species to the length of two or three feet. In certain of the Hypnums, particularly the H. dendroides and H. alopecurum, may be found miniatures of every tree in an arboretum. Comparing the flowering plants with the mosses, we shall find something corresponding in each genus of the one to something in each order or family of the other. Thus while every kingdom starts on a platform of its own, and enjoys a perfection no less peculiarly its own, it anticipates the kingdom that is above it, and though the several perfections are unlike, there is thus a fine harmony between them.

Mosses directly serve very few purposes in the economy of man. They are often employed for packing articles, for which they are admirably adapted. Linnæus informs us that the Swedish peasantry fill up the spaces between the chimney and the walls in their houses with a particular kind, which prevents the action of the fire by the exclusion of air, viz., the *Fontinalis antipyretica* or great water-moss, which forms enormous masses a foot or more long, floating in rivers and stagnant water. Another species is sometimes employed in the manufacture of mats and brooms. The bog-moss supplies materials for mattresses. The

Laplanders use it instead of clothes for their new-
born babes, packing their cradles firmly with it ;
and in seasons of scarcity it enters into the composition of their bread.   The dense fork-moss, when
twisted, is used by the Esquimaux for lamp-wicks,
a purpose which it very inadequately performs.
But this is about all that can be said of their value
to man.   In the economy of Nature, however, they
are extremely useful.   They contribute to the
diffusion and preservation of vegetable life, both
by the soil which their decay supplies, and by the
shelter which they afford to the roots of trees and
plants in very hot or very cold weather.   Peat is
almost entirely composed of mosses.   This substance is usually found in great basin-shaped hollows, or valleys among the hills, formerly covered
with indigenous forests of birch, alder, and hazel,
or with the waters of a mountain lake.   In the former case, the rotting of the fallen trees produced
a rich black mould where mosses luxuriated ;
these mosses acted like sponges, and absorbed the
moisture from the atmosphere, and retained the
rains when they fell, forming shallow marshes
around the fallen trees.   More mosses were developed by this moisture, and more moisture was
accumulated by these mosses; and thus the mutual
process went on, one layer of moss decaying in
its lower parts, and increasing by additions to

its tops—the dead giving birth to the living—until
at last the fallen trees were completely entombed,
and a stratum of upwards of twenty feet of solid
peat, in some instances, deposited above them.
When, on the other hand, the basin-shaped hollows were originally occupied by lakes, the *Sphagnum* or bog-moss abounded in the waters, and
spread so extensively, even from great depths, as
through course of time to transform the lakes into
quaking bogs, which, by the accumulation of drift,
dust, and rubbish, and the decay of the original
plants and the formation of new, became ultimately compressed into solid peat, covered upon the
surface with heather, or a green vesture of grass or
moss. The *Sphagnum* or bog-moss by which this
great change was effected is of a singularly pale,
almost snowy-white colour, a peculiarity exceedingly rare among plants, and sometimes attains a
length of six or seven feet in deep water ; its large
air-cells imparting the necessary buoyancy to it.
Its structure is in many respects different from
that of all other mosses. Its branches are
fasciculate and disposed around the stem in
spirals ; it has no roots whatever, but floats unattached in an upright position in the water ;
its cell-walls are perforated, and the leaf-cells contain a well-developed spiral ; while the stem is composed of tissue, which, under the microscope, bears

E

a close resemblance to the glandular structure of the stems of coniferous trees. The seed-vessel rises among the leaves on a peduncle resembling a fruit-stalk, and bursts in the centre, the lid flying off when the seed is ripe with considerable force. It is extensively distributed in temperate regions, being almost unknown in the tropics, where the peat is formed by the decomposition of shrubby plants like the common heather. The peat of Tierra. del Fuego, the Falkland Islands, and the Galapagos Archipelago, is composed of this bog-moss. It is geographically interesting to find a species, *Sphagnum Austini*, not rare in North America, growing on boggy moors and forming large hummocks sometimes two feet above the surrounding level, in the Island of Lewis, one of the Hebrides, the only place in which it has yet been found in this country. We may be able to form some idea of the vast importance of the Sphagna, when we consider that peat-bogs occupy a tenth part of the whole of Ireland, and furnish in the Highlands of Scotland the largest proportion of the fuel consumed by the inhabitants. It is a singular fact that we owe our coals to the carbonized remains of ferns and their allies ; and our peats to the decomposed tissues of mosses—two of the most useful and indispensable materials in our social economy to two of the humblest families in the vegetable kingdom. How true it is, that things which we are apt to despise or overlook on account of their minuteness and apparent insignificance, are not only full of lessons of beauty and wisdom, but are also made the means, in the hands of a kind Providence, of the greatest good to His creatures !

HEPATICAE OR SCALE MOSSES.

The plants whose peculiarities have been described in the preceding pages are called Urn Mosses, their fructification being urn-shaped, furnished with teeth, and closed with a lid. There is another large class called Scale-Mosses, so closely allied to the true mosses that they are frequently confounded even by an educated eye. They are united together by connecting links. On Alpine rocks a genus of dark-brown, almost black mosses called *Andræa* occurs in wide-spreading tufted masses, which seems to combine the characters of both the urn and scale mosses. Its.capsule opens by four valves, thus resembling the four-valved spore-case of the *Jungermanniæ* or scale-mosses ; while it has the persistent lid, the columella, and the characteristic foliage and habit of the true mosses. The two divisions of mosses also shade into each other in the broad, entire, nerveless leaves of the

shining *Hookeria* (*H. lucens*), one of the loveliest
of our native mosses occurring sparingly and
locally on moist banks, in shady woods, and in the
entire nerveless flat leaves and stems of the Flat
Feather-moss (*Hypnum complanatum*), one of the
commonest of all mosses on the trunks of trees.
There are upwards of a hundred species of scale
mosses indigenous to Great Britain and Ireland,
some of which are so small as to be scarcely·

FIG. 2.—JUNGERMANNIA COMPLANATA.

visible, and others much larger than any of the
true mosses. With the exception of a few prominent species, which are found in every moist wood
and on every shady rock, they are somewhat local
and limited in their distribution, many of them
being remarkably rare, and confined to remote
and isolated localities. The greatest number of
species occurs in the tropics ; and nowhere do
they luxuriate so much as in the dark woods and

mountain ravines of New Zealand. Some of them grow in the bleakest spots in the world, and are to be found even at a higher altitude than the urn-mosses on the great mountain ranges of the globe. They form the faintest tint of green on the edges of glaciers, and on the bare storm-seamed ridges of the Alps and Andes, where not a tuft of moss or a trace of other vegetation can be seen ; and this almost imperceptible film of verdure, when cleansed from the earth and moistened with water, presents under the microscope the most beautiful appearance.

The peculiarities of these plants are so remarkable and interesting that they deserve more than a passing notice. As a rule, to which however there are a good many exceptions, they do not grow upright in tufts like the mosses, but have a flat, creeping, lichen-like habit, spreading over rocks and trees in closely-applied circles which radiate from a common centre. The whole typical plant is like a series or necklace of roundish flat imbricated scales, several of which branch from a common point in the middle. The leaves, unlike those of the mosses, are entirely destitute of a central nerve, for what is called the nervure in the membranous or leafy species is nothing more than the stalk itself, on the edges of which the leaves are fastened together in such a manner as

to form apparently a continuous whole. They are
disposed either in a spiral which turns from left to
right or from right to left. They overlap each
other in two ways ; either each leaf covers with its
lower edge a little of the leaf below it, in which
case the leaves are called *succubous*, or each leaf
overlaps a little of the base of the leaf above it, in
which case the leaves are called *incubous*. In their
shape there is a marvellous diversity, being frequently deeply toothed or bi-lobed ; and the arrangement and form of the cells is so exquisitely
beautiful in almost all the species, that no more
pleasing objects can be mounted for the microscope. The grains of chlorophyll often exhibit in
them, as in the leaves of the true mosses, apparently spontaneous movements under the influence
of light. If kept in the dark for several days the
cells present the appearance of a green net-work,
between the meshes of which is a clear transparent
ground. All the grains of chlorophyll are applied to
the walls which separate the cells from one another.
If placed again in the sunshine, or even under the
influence of artificial light, the grains change their
position from the lateral to the superficial walls—
where, however, they do not remain absolutely im-
moveable, but continually approach and separate
from one another for upwards of a quarter of an
hour. If again darkened they leave their new posi-

tion and return to the lateral walls of the cells.
These plasmic movements are exceedingly curious
and interesting.

The lowest forms of scale-mosses consist simply
of a patch of green membrane spreading over
the ground composed of a single or double
layer of cells containing chlorophyll. Higher
types have a more definite outward appearance, a greater complexity of internal structure,
and possess a skin investing both surfaces—the
upper portion of the frond containing stomata or
breathing pores like the leaves of flowering-plants.
From the slight groove which runs along the middle
line on the upper surface of some species, and
which looks like a mid-nerve, arise minute leaf-
like bodies called amphigastria, which resemble
the stipules of Phanerogamous plants ; while from
the projecting rib on the lower surface corresponding to the groove on the upper, arise numerous
radicles or rootlets, although in some species they
are scattered indiscriminately over the whole under-
side. Their substance is very loosely cellular,
easily reviving, after being dried, by the application
of moisture. The species that have stomata or
breathing pores, however, when once dried, revive
very slowly and imperfectly ; being in this respect
analogous to flowering plants. Their colour varies
from a pale white to the darkest green and the

deepest and most brilliant red and purple ; sea-
green, however, being the prevailing hue. The
fruit-vessel is as interesting and suggestive of
marvellous reflection as that of the urn-mosses.
It is generally supported on a very delicate silvery
stem ; and is at first round and of a dark olive-
green colour, gradually splitting as it becomes
ripe into two or four valves, which bear a superficial
resemblance to the calyx or corolla of flowering
plants.   In the centre of this calyx-like organ may
be seen a tuft of delicate straw-coloured hairs or
filaments called elaters, which look like floss silk,
with the spores or seeds in the form of minute
yellow dust intermingled.   These filaments are
spiral, highly elastic, and hygrometrical, twisting
and writhing even upon the field of the microscope ;
and like the spring-like ring round the fruit-vessel
of the fern, serve by their coiling and uncoiling, in
certain states of the surrounding atmosphere, to
scatter abroad, even to a considerable distance,
the powdery seeds imbedded among them.   This
is a very curious and wonderful piece of mechanism,
and highly deserving of microscopical examination.
The formation of the fruit-vessel is preceded by
the development of antheridia and pistillidia, which
demonstrate the existence of distinct sexes in
these plants.   They are produced on different
places in different species, being imbedded in the

substance of the frond, or occurring free in the axils of the leaves, or immersed in special stalked receptacles. Besides this mode of reproduction, the scale-mosses are propagated by *gemmae* or little cellular nodules, which in one genus are produced in elegant cups with a toothed margin growing on the upper surface of the frond ; by *innovations* or new lobes growing out from the margins of the old fronds ; by buds in the axils of leaves ; and by confervoid branches sent out from the stem. The spores that are produced by the normal mode of reproduction have in most of the species a double coat, and germinate by protruding pouch-like processes from which the new fronds or leafy stems arise.

The Hepaticae or scale-mosses may be divided into two groups, consisting of those species in which the vegetation is frondose, that is, in which leaf and stem are confounded, and of those in which the vegetation is foliaceous, that is, in which leaves and stem are distinct. One genus of the frondose group called Riccia or crystal-wort floats on the surface of stagnant waters, and bears a superficial resemblance to the common duck-weed. The fronds, which are exceedingly delicate cellular leaf-like structures, are destitute of radicles when growing on the surface of ponds and ditches ; but if the water be removed by evaporation or draining, or the plant thrown on the soil at the margin, they become smaller, and fasten themselves firmly to the ground by numerous fibrous rootlets,—a beautiful example of the ease with which these humble plants accommodate themselves to altered circumstances. They have many air-passages between the cells, which enable them to float on the water. The under surface is covered, to a greater or less extent, with thin scales, which form most beautiful microscopic objects when treated with different chemical tests, from their transparency and variety of colouring. One ally of this genus, called Riella, differs widely from the rest of the tribe in its erect, moss-like habit. It grows on the margins of ponds, streams, and lakes in Algiers and Sardinia, and perfects its fruit when submerged. It is quite a botanical curiosity, presenting a whorled appearance, not unlike the common spiral shells of the sea-shore. Each individual consists of a central stem, round which a distinct leaf or wing is wound in the form of a screw or continuous spiral. On the edge of this wing, towards the summit of the male plant, the antheridia are developed ; while in the female the fruit clusters on the stem between the whorls. An example of this beautiful genus is very common on moist garden paths and on the mould of pots in the green-house and stove. It forms little

star-like tufts radiating from the centre, and presenting a remarkably pellucid crystalline appearance. The fruit-vessel is immersed in the frond,
and has no elaters. It never bursts regularly, but
emits the spores only by decay.

The most interesting of all the frondose group of
scale-mosses, however, is the common Marchantia
or Liverwort (*Marchantia polymorpha*, Fig. 3). It
is very common, creeping in large, dull, dark-

FIG. 3.—MARCHANTIA POLYMORPHA.

green patches over rocks in very moist and shady
situations, such as the banks of a densely-wooded
stream in a deep narrow glen, or the sides of
rivers and fountains. It may often be seen also
on the moist walls of hot-houses, and in the pots
and tubs. It adheres closely to rocks, which it
sometimes completely covers with its imbricated
fronds, by the numerous white downy radicles
with which the under-surface is covered. Its
fronds are flat, about three inches long, and from

half-an-inch to an inch wide, and are variously
divided into obtuse lobes. Their texture is
membranaceous and strikingly cellular. Their
upper surface is most beautifully reticulated and
covered with numerous minute lozenge-like scales,
with a little dot-like pore or puncture in the
centre, analogous to the stomates or breathing-
pores of flowering plants. The fructification is
very singular, resembling a forest of little mushrooms rising from the leaves ; each dividing at the
top into eight or ten green rays, and having as
many little brown purses placed alternately
between them. Each of these purses has a valve
which opens generally in July, and contains
within it four or five florets, from the centre of
which rises a single funnel-shaped filament,
covered with a yellow powder affixed to the
elaters or elastic spiral hairs previously alluded to,
exhibiting highly developed spiral bands. Besides
this ordinary male and female stalked receptacle,
sterile as well as fertile individuals are provided
at all seasons of the year with cup-like bodies,
growing on various parts of the upper surface of
the frond, and of the same texture as the frond
itself. These bodies seem to indicate an approach
to the calyx and corolla of the flowering plants.
They contain in their interior several lentil-shaped
membranaceous bodies of a reticulated structure,

equivalent to buds, which frequently throw out rootlets before leaving their receptacles, and, striking root on the spots where they happen to fall, in time become perfect fronds. There is no more pleasing and profitable study to the young botanist than the examination of the highly curious structure and complex system of fructification peculiar to this plant. It is interesting also on account of its associations. Under the name of *Hepatica officinarum*, it was employed by the ancient herbalists, from its resemblance to the reticulated structure of the liver, as a cure for all diseases affecting that organ. It is still used as a popular remedy for jaundice and other maladies in some parts of England; but its virtues are, in all likelihood, entirely imaginary. Hoffmann and Willemet, in their elaborate treatise upon the uses of lichens, state regarding it, 'Cette plante est amère, aromatique, abstersive, vulnéraire, sudorifique, apéritive. On prescrit l'Hépatique en apozème, à la dose d'une poignée pour l'homme, et de deux ou trois pour les animaux.' The bruised fronds of some species are singularly fragrant, resembling bergamot.

The second or foliaceous group of scale-mosses, in which the leaves and stem are distinct, is called *Jungermanniæ*, and contains by far the largest number of species, and the richest variety of form and colour. On either side of the thread-like stem arise in a more or less oblique position the membranous overlapping leaves ; while the fruit-vessel springs from the end of the stem, and is produced upon little silvery foot-stalks. It bursts into four valves, and when fully expanded spreads out into the form of a cross. All the species of this group may be known at once, and distinguished from the true mosses by the imbricated and peculiarly flattened appearance of the leaves. One of the commonest representatives of the *Jungermanniæ* is the *J. dilatata*, which is found in every wood covering the trunk of almost every tree with its rich chocolate-coloured masses. If it does exhibit any preference, it is for the smooth stem of the poplar and the mountain-ash, to which it imparts a singularly picturesque appearance. A kindred form often confounded with it, the *J. Tamarisci*, spreads in large loose tufts upon the ground, and over low bushes in sub-alpine countries, and is of a glossier texture and richer colour, brightening into pale maroon and amber at the edges. It is exceedingly beautiful even to the naked eye. Contrasting strikingly with these brown species, are the pale green orbicular patches closely pressed to the bark of trees, of the equally common *J. complanata*, which fruits throughout the year. One of the largest species, growing from three to six inches in length, with ascending branched stems and large round pellucid leaves, is the spleenwort Jungermannia (*J. asplenioides*), which is exceedingly common in moist woods, on shady banks, and among rocks. Equally large and growing in great tufts of a rich purple or even crimson colour, in the beds of mountain streams or on moist moors, is the shell-leaved Jungermannia (*J. cochleariformis*). It is as lovely in texture as it is in hue. But the most beautiful of all the species is the *J. tomentella*. It is a very peculiar plant, and like no other European species. Its leaves, which are peculiarly pale in colour, and so crowded and cut into fine capillary interwoven segments, that the whole has almost the texture of sponge or flannel. It is found in great abundance where it does occur, although it is somewhat local and restricted in its distribution. In sub-alpine woods in the Highlands of Scotland it is by no means rare, and when it grows in great masses protruding its upper lobes, tier above tier on wet rocks beside waterfalls, it forms one of the loveliest spectacles upon which the eye of the lover of Nature can gaze, and which one would go far to see.

LYCOPODS OR CLUB-MOSSES.

There is a class of plants whose external appearance and mode of growth would indicate that they belong to the tribe under review, but whose structure and functions are so different, that they are commonly supposed to bear a closer analogy to the ferns. They occupy an intermediate position, and form a connecting link between ferns and mosses; I allude to the Lycopods or club-mosses. They are usually found in bleak, bare, exposed situations in all parts of the world, and sometimes attain a large size ; forsaking the creeping habit peculiar to the family, and becoming slightly arborescent in tropical countries, particularly New Zealand, rivalling in rank luxuriance the smaller shrubs of the forest. The British representatives of the class are comparatively small plants, with the exception, perhaps, of the commonest species (*Lycopodium clavatum*, Fig. 4), which creeps along the ground among the heather on the moorlands, and sends out runners or creeping stems in all directions to the length of several yards, which take a firm hold of the soil by means of long, tough, wiry roots on their under-surface. The smallest species is the marsh club-moss (*Lycopodium inundatum*), which grows upright in little tufts at the edge of streamlets, or in marshy hollows among the hills where it is almost wholly concealed by the surrounding bog-mosses. In this country the lycopods are all alpine or sub-alpine ; one species (Fig. 5) ascending to the highest summits of the British mountains, where it grows in large rigid tufts amid the *débris* of rocks, and another

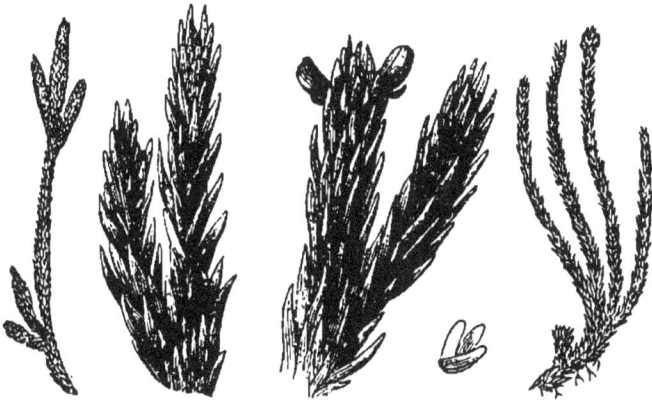

FIG. 4.
LYCOPODIUM
CLAVATUM.

FIG. 5.
LYCOPODIUM
SELAGO.

FIG. 6.
LYCOPODIUM
ALPINUM.

(Fig. 6) trailing in long wreaths over the bare mossy shoulders of the Highland hills, sending up at short intervals from the bare, whitish, procumbent stems, palm-shaped tufts of very hard foliage, very like that of the savine. The loveliest of the British species is the *L. selaginoides*, which looks like a moss as it creeps among mosses on sub-alpine banks. It is a slender and delicate species ; its fructifying spike being of a

F

bright golden colour, and of a glossy, almost pellucid texture. In other parts of the world, however, Lycopods grow on the low grounds in the woods and other warm, humid situations, adding to the picturesqueness and beauty of the sylvan scenery. One species, the Tmesipteris, remarkable for its pendulous habit and very broad leaves, hangs down in long trailing wreaths from the trunks of tree-ferns, in South America and New Zealand. In the little island of St. Paul, isolated from the rest of the world in the Indian Ocean, thousands of miles from any friendly shore, there occurs a beautiful species (*L. cernum*), the presence of which in that remote locality is a puzzle to the student of geographical botany. This island is situated in the temperate zone, while the normal range of this plant is exclusively within the tropics. As, however, the Island is volcanic, and contains numerous hot springs, which diffuse considerable warmth around, this circumstance may account for the presence of the lycopod, especially as it also occurs, far out of its proper range, about the warm springs of the Azores. Luxuriating in beautiful tufts amid the barren tufa of this lonely island, it is a welcome and refreshing sight to the voyager on the way to Australia, tired of the monotony of the sea, and yearning for mother earth. Like himself, a stranger in a strange land, it often reminds the emigrant of the brown moorlands of his native country, where he used to gather the trailing wreaths of the fox-fetters to bind around his cap in the sunny days of youth. One remarkable species (*Lycopodium squamatum*), which grows in the arid deserts of central South America, among aloes and cactuses, is possessed of singular hygrometric properties. In the dry season, when every particle of moisture is extracted from the soil, it rolls itself up into a ball, like the young frond of a fern before it is unfolded, and unrolls during the wet season, recovering its green colour and spreading itself out flatly on the soil like a branch of arbor-vitæ, with its former vigour and freshness. There are several other species in Mexico and Brazil which also curl up and contract into a ball in the dry season, and losing their hold upon the soil, are blown across the plains by the violent equinoctial gales that prevail at the time, like the Anastatica or "Rose of Jericho" of Palestine. They are often brought to this country and pre- served by the curious under the name of the "Resurrection Plant," who think them still alive because they expand when placed in water. A singular phenomenon has been observed in a species of Selaginella cultivated in Kew gardens, called specifically from this circumstance *mirabilis*.

In the morning the fronds are green, but as the day advances they become pale, recovering gradually their colour by the following day. Dr. Hooker has observed that in their pale condition the chlorophyll of the cells of the leaves is contracted into a little pellet. This phenomenon is of the same nature as that which has been already described in connexion with the chlorophyll movements in the cells of mosses and hepaticae under the stimulus of light. The colour of some Lycopods is of a bluish metallic tinge, and seems to depend upon the effect of the different arrangement of the chlorophyll in the cells of the leaves upon the light.

The club-mosses bear in the axils of their leaves minute round or kidney-shaped cases of a bright yellow colour, which form the receptacles of their dust-like seed. Some species have little cone-like spikes at the tips of their branches, under the scales of which, as in the pine tribe, lurk the reproductive embryos. In the common club-moss these spikes are two-pronged, and of a whitish colour, while the seed is highly inflammable, and was formerly employed to produce artificial lightning on the stage, by being blown through a tube and ignited. It is equally remarkable for the way in which it repels moisture, and for this reason it is employed by druggists in the manufacture of pills. It originates independently of any reproductive organs or fertilizing influence. Indeed it is these seeds in germination which develop the structure upon which the fertilizing organ, and the organ to be fertilized, are situated. The stems are perennial, and consist of a mass of thick walled, often dotted cells, enclosing one or more bundles of scalariform tissue, which send off branches to every leaf and bud. Among these bundles may be seen elongated cells, distinctly reticulated. This kind of tissue indicates a close relation to the ferns, and justifies the position in which they are usually placed by systematists. New fruit-axils are formed year after year, bearing their new cluster of seeds independent altogether of any fertilizing organs, such as antheridia and archegonia. The club-mosses are all very graceful and beautiful plants. The Spanish moss (*Lycopodium denticulatum*) is a great ornament to conservatories and hot-houses, where it conceals with its luxuriant drapery the mould in the pots, and keeps the roots of the plants moist. Nothing can be lovelier or more elegant than a basket of orchids in full flower, with clusters of this moss drooping in careless grace from its sides. The common club-moss of our moors is often gathered by the peasantry to festoon the ornaments of their mantelpieces;

while wreaths of it are collected from the woods
of Balmoral, where it grows in abundance, to
grace the royal table. At the Lakes of Killarney
the Irish tourist is compelled to purchase from
his pertinacious followers the "Blessed Fir" as the
people call the Lycopodium Selago. All the
species of lycopods are possessed of poisonous, or
at least questionable properties. The *L. cathar-*
ticum has been administered as a strong cathartic.
In the Highlands they are employed with alum
as a mordant to fix the native dyes in the manufacture of tartan, while they are said themselves
to produce a blue tint.

Lycopods may be said to present the highest
type of cryptogamic vegetation, the highest limit
capable of being reached by flowerless plants.
Indeed, they bear a very close affinity to
coniferous trees, a resemblance which even the
Irish peasantry have recognised in the name they
have given to the Killarney species. The *Lyco-*
podium dendroides of North America resembles at
a distance young spruce firs, being similarly shaped
and of a lively green colour. This affinity, though
indicated by very curious resemblances, is, however, strictly analogical. The gap between the two
great orders of plants is too wide to be overleaped
by a sudden transition. There is a resemblance
in external form, habit, and fructification; the

leaves are in both cases linear; the seeds are in both cases produced from cones or spikes; the formation of the archegonia and embryonic pods of the one is similar to that of the corpuscles and embryo in the other, but in these points the likeness begins and ends. The resemblance which we see between the Lycopod and the conifer, is like that which exists between the cucumber and the passion-flower, the water-lily and the poppy and magnolia. In the Compositæ, the largest of all Phanerogamous orders, the habit of almost every other order of the vegetable kingdom crops up again. Every platform of plants is found in close analogy with every other platform. There is nothing in Exogens which we do not find in Endogens ; nothing in flowering plants which we do not find among flowerless plants. In the strange Brazilian family of Podo-stemas we see liver-worts and scale-mosses in flower ; while in those curious trees of Australia, the Casuarinas, reappear the leafless branches and singular joints of the Equisetums or Horse Tails of our marshes. But we must remember that often where there is the greatest amount of appa-rent affinity, there is the least real affinity; that in judging the value of such mimicries, echoes, or resemblances " like is an ill mark." There is no true homology, but a mere analogy which is often

seen to harmonize the most dissimilar works of
nature, as if to show that they proceeded from
the same creating Hand. There may be a gradual
transition from one class of plants to another, and
certain characters may be common to two
families ; but still there are definite groups in
nature, and typical characters belonging to
plants, which will for ever keep them distinct and
isolated, as illustrations of the infinite variety of
the Divine works.

The first pages of the earth's history reveal
to us very extraordinary facts with relation to
members and allies of the moss tribe. The club-
mosses, in particular, at a former period, seem to
have played a more important part, or to have
found conditions more suitable to their luxuriant
development than is the case at the present day.
The two or three hundred species at present
existing are the mere remnant of a once magnificent group. Some of them are stated to have
formed lofty trees eighty feet high, with a proportionate diameter of trunk. They are among the
most ancient of all plants. The oldest land-plant
yet known is supposed to be a species of lycopodium closely resembling the common species of our
moors. In the upper beds of the Upper Silurian
rocks, they are almost the only terrestrial plants
yet found. In the lower Old Red Sandstone they

also abounded ; while they occupied a considerable space in the Oolitic vegetation. But it is in
the Coal-measures that they seem to have attained
their utmost size and luxuriance, sigillaria, lepido-
dendron, etc., being now considered by competent
botanists to be highly-developed lycopodia. Along
with ferns, they covered the whole earth from
Melville Island in the Arctic regions to the Ultima
Thule of the Southern Ocean, with rank majestic
forests of a uniform dull green hue. The
numerous coal-seams and inflammable shale
found in almost every part of the world, form
but a small portion of their remains. "Between
the time of the ancient lycopodite found in the
flagstone of Orkney," says Hugh Miller, "and
those of the existing club-moss that now scatters
its light spores by millions over the dead and
blackened remains of its remote predecessor,
many creations must have intervened, and many
a prodigy of the vegetable world appeared,
especially in the earlier and middle periods,—
Sigillaria, Favularia, Knorria, and Ulodendron,
that have had no representatives in the floras of
later times ; and yet here, flanking the immense
scale at both its ends, do we find plants of so
nearly the same form and type that it demands
a careful survey to distinguish their points of
difference."

CHAPTER II.

LICHENS.

O most minds the title of this chapter may suggest no idea of importance. Flowers they love, for they are linked with childhood's recollections of sunshine and mirth, and mingle with the hallowed memories of the dead, and of the scenes amid which they are laid. Ferns they admire as they cluster in the forest shade, gracefully bend down to see their own forms in the mossy spring, or wave from some rugged crag their delicate fronds in the breeze of summer. Mosses they allow to be lovely, as they repose their languid limbs in the sultry noonday, on the woodland banks wreathed in dreamy-looking shadows, to which these tiny plants lend their all of softness and beauty. But the lowly lichens they pass by with indifference, regarding them only as inorganic discolorations and weather-stains on the trees and rocks where they repose. And yet they too are interesting, both as regards their history and their uses; as interesting as many plants which occupy a far higher position in the ranks of vegetation. Uninviting and apparently lifeless although their external aspect may appear, they are found, when subjected to the microscope, to have their own peculiar beauties and wonders. Simple as is their construction, being entirely composed of an aggregate of minute cells united together in various ways by intercellular matter, and completely destitute of stems, leaves, and all those parts which enter into our ideas of perfect plants, yet by a wonderful compensation they are so extensively diversified in their form and appearance, as to present to the student of nature a field for his inquiry, as wide and wondrous as the display of green foliage and blossoms of every hue which glow in the summer sun.

To the landscape painter, intent upon seeking materials for the foregrounds of his sketches, lichens possess a special interest. Through their instrumentality the miserable hovel, with its rough unmortared walls, becomes a charming and romantic object. The old dike by the wayside, commonplace and disagreeable although it may look when newly constructed, becomes a pleasing feature in the landscape when garnished with the

grey rosettes, eccentric patches, and nebulæ of
the lichens ; and the rude rugged rock acquires an
additional wildness and picturesqueness through
the affluent display of these plants. Along with
the wallflower and the ivy, they decorate the
mouldering ruin, and harmonize its otherwise
haggard and discordant features, by their subdued
and varied colouring, with the gentler forms and
the softer tone of the scenery around. Thus
nature takes back into her bosom the falling works
of human skill and power, and luxuriantly adorns
them with her living garniture of beauty; and
these softening stains with which she touches the
rude, stern masses she disjoins, have their value
in the composition not simply on account of the
pleasure they afford to the eye by the mere tints
of a painter's palette, but also and chiefly on
account of the meaning they suggest through the
eye to the mind as the genuine and expressive
colouring of time. To the trees of the forest
lichens impart a singularly aged and venerable appearance which irresistibly commands our homage,
and leads our thoughts far back over the dim path of
years to the memories of primitive times. So abundant are they in the Highland woods, that every tree
is covered with their long white streaming tufts,
which look, on the green tassel-laden branches,
and among the fringy, waving hollows of the

pyramid-like foliage, like the snowy blossoms of
some unknown fruit-tree. It is impossible to enter
a pine forest adorned with a profusion of these
curious plants, without admiring the wild and
picturesque appearance which it presents. The
hoary trees seem like an assembly of aged bearded
Druids, metamorphosed by some awful spell while
in the act of worshipping their mysterious deity ;
while the feelings of solemn awe and reverence with
which we regard them are rendered more intense
and overpowering by the dread silence, the utter
solitude that reigns around—a silence broken only
by the low, deep, sybilline sigh of the wind among
the tree-tops ; the faint crackling sound of the
falling pine-cones ; or perchance, at rare intervals,
the wild, melancholy cries of some little wander-
ing bird afraid to find itself alone in such a dreary
place, multiplied with startling distinctness through
the forest as they pass along from echo to echo.
Perhaps a red-deer stands gazing at you, with
large inquiring eyes, at the end of a long vista
between the red trunks of the trees ; but as you gaze,
it glides away into a deeper solitude as noiselessly
and as mysteriously as it came ; and the very
sunbeams, that elsewhere dance and sport with
the wavering shadows, and chase each other in
long links of golden light over the mossy sward,
creep through the dense canopy overhead, and

down the lichened trunks slowly and hesitatingly, as though, like children who stand at the mouth of some grim yawning cavern, they longed yet dreaded to enter. How applicable to this weird scene is the graphic description of an American forest, with which Longfellow opens his beautiful poem of "Evangeline"—

> "This is the forest primeval. The murmuring pines and the hemlocks,
>
> Bearded with moss, and in garments green, indistinct in the twilight,
>
> Stand like Druids of old, with voices sad and prophetic;
>
> Stand like harpers hoar, with beards that rest on their bosoms."

We are more indebted to the humble lichens for the charming romance of our sylvan scenery than we imagine ; for we are apt to overlook the minute plants by which much of the effect is produced. All who have any taste or poetical feeling admire the conspicuous beauties of a wood—the clouds of green foliage overhead, the endless ramifications of the branches, the massiveness and elegance of the trunks, and the softness and richness of the grassy carpet underneath ; but there are few, comparatively, who pay any attention to those minute varieties of tint and form contributed by the lower orders of vegetation—the starry flower, the plumy fern, or the umbrella-like fungus upon the ground, and the clustered moss and trailing lichen upon the tree ; and yet it is with these small and apparently insignificant objects that nature shades the picture, balances and contrasts the colouring, clothes the nakedness, and softens down the irregularities and deformities of the whole scene, which would otherwise be stiff and hard as a forest-piece painted by a Chinese artist.

Lichens are exceedingly diversified in their form, appearance, and texture. About five hundred different kinds have been found in Great Britain alone, while upwards of three thousand species have been discovered in different parts of the world by the zealous researches of naturalists. In their very simplest rudimentary forms, they consist apparently of nothing more than a collection of powdery granules, so minute that the figure of each is scarcely distinguishable, and so dry and utterly destitute of organization that it is difficult to believe that any vitality exists in them. Some of these form ink-like stains on the smooth tops of posts and felled trees ; others are sprinkled like flower of brimstone or whiting over shady rocks and withered tufts of moss ; while a third species is familiar to every one, as covering with a bright green incrustation the trunks and boughs of trees in the squares and suburbs of smoky towns, where the air is so impure as to forbid the growth of all other·vegetation. It also creeps over the grotesque figures and elaborate carving on the roofs and pillars of Roslin .Chapel, near Edinburgh, and gives to the whole an exquisitely beautiful and romantic appearance. One species, the *Lepraria Jolithus*, is associated with many a superstitious legend. Linnæus, in his journal of a tour through Œland and East Gothland, thus alludes to it :—" Everywhere near the road I saw stones covered with a blood-red pigment, which on being rubbed turned into a light yellow, and diffused a smell of violets, whence they have obtained the name of violet stones; though, indeed, the stone itself has no smell at all, but only the moss with which it is dyed." At Holywell, in North Wales, the stones are covered with this curious lichen, which gives them the appearance of being stained with blood ; and of course the peasantry in the neighbourhood allege, that it is the ineffaceable blood which dropped from St. Winifred's head, when she suffered martyrdom on that sacred spot. A higher order of lichens (*Bæomyces*) is furnished, besides this powdery crust, with solid, fleshy, club-shaped fructification like a minute pink fungus ; while a singularly beautiful genus (*Calicium*), usually of a very vivid yellow colour, spreading in indefinite patches over oaks and firs, is provided with capsules somewhat like those of the mosses. These capsules, though thickly scattered over the crust, are so minute as to be scarcely distinguishable by the naked eye, but under the microscope they present a truly lovely appearance. They are cup or urn-shaped, of a coal-black colour, and supported by a slender stalk about the thickness of a horse-hair. At an early stage they are covered with a very delicate veil, which stretches completely over their mouth; but this soon vanishes, and exposes to view a mass of black or brown seeds, like the ovule in an acorn, which the slightest touch of the tiniest insect's wing can dislodge, and send away on the breeze in search of a habitat for another colony.

Most of the crustaceous lichens are merely grey filmy patches inseparable from their growing places, indefinitely spreading, or bounded by a narrow dark border, which always intervenes to separate them when two species closely approximate, and studded all over with black, brown, or red tubercles. The foliaceous species are usually round rosettes of various colours, attached by dense black fibres all over their under-surface, or by a single knot-like root in the centre. Some are dry and membranaceous; while others are gelatinous and pulpy, like aërial sea-weeds left exposed on island rocks by the retiring waves of an extinct ocean. Some are lobed with woolly

G

veins underneath; and others reticulated above, and furnished with little cavities or holes on the under-surface. The higher orders of lichens, though destitute of anything resembling vascular tissue, exhibit considerable complexity of structure. Some are shrubby, and tufted, with stem and branches, like miniature trees; others bear a strong resemblance to the corallines of our sea-shores; while a third class, "the green-fringed cup-moss with the scarlet tip," as Crabbe calls it, is exceedingly graceful, growing in clusters beside the black peat-moss or under the heather tuft,

> " And, Hebe-like, upholding
> Its cups with dewy offerings to the sun."

As an illustration of the extraordinary appearances which lichens occasionally present, I may describe the *Opegrapha* or written lichen (Fig. 7), perhaps the most curious and remarkable member of this strange tribe. In her cactuses and orchids sportive nature often displays a ludicrous resemblance to insects, birds, animals, and even the "human face and form divine;" but this is one of the few instances in which she has condescended to imitate in her vegetable productions the written language of man. A cryptogam is in this case a cryptogram! The crust of this curious autograph of nature is a mere white tartareous film of indefinite extent, sometimes bounded by a faint line of black like a mourning letter. It spreads over the bark of trees, particularly the beech, the hazel, and the ash. On the birch-tree—whose smooth, snow-white, vellum-like bark seems designed by nature for the inscription of lovers' names and magic incantations—it may often be seen covering the whole trunk. The fructification consists of long wavy black lines, sometimes parallel like Runic inscriptions ; sometimes arrow-

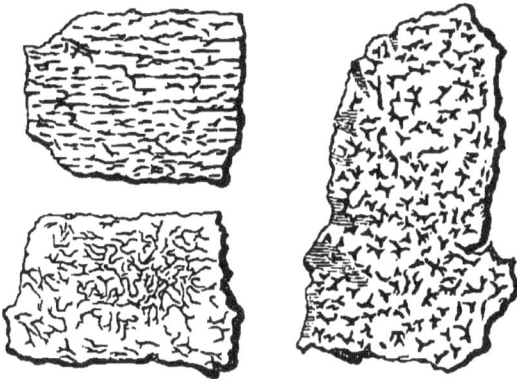

FIG. 7.—OPEGRAPHA SCRIPTA.

headed, like the cuneiform characters engraved upon the monumental stones of Persepolis and Assyria ; and sometimes gathered together in groups and clusters, bearing a strong resemblance to Hebrew, Arabic, or Chinese letters.

In that well-known and interesting work, *Travels in Tartary, Thibet, and China,* by the French Lazarists Huc and Gabet, there is a long description of a very remarkable phenomenon called the

"Tree of Ten Thousand Images," found by them near the town of Koumboum in Thibet. For the sake of those who may not have access to the original work, I shall quote the description entire. "At the foot of the mountain on which the Lamasery stands, and not far from the principal Buddhist temple, is a great square enclosure, formed by brick walls. Upon entering this, we were able to examine at leisure this marvellous tree, some of the branches of which had already manifested themselves above the wall. Our eyes were first directed with earnest curiosity to the leaves, and we were filled with an absolute con- sternation of astonishment at finding that in point of fact, there were upon each of the leaves well formed Thibetian characters, all of a green colour, some darker, some lighter than the tree itself. Our first impression was a suspicion of fraud on the part of the Lamas; but after a minute examina- tion of every detail, we could not discover the least deception. The characters all appeared to us portions of the leaf itself, equally with its veins and nerves ; the position was not the same in all ; in one leaf they would be at the top, in another in the middle, in a third at the base or at the side ; the younger leaves represented the characters only in a partial state of formation. The bark of the tree, and its branches—which resemble that of the plane-tree—are also covered with these characters. When you remove a piece of old bark, the young bark under it exhibits the indistinct outlines of characters in a germinating state, and what is very singular, these new characters are not unfrequently different from those which they replace. We examined everything with the closest attention, in order to detect some trace of trickery, but we could discern nothing of the sort ; and the perspiration actually trickled down our faces under the influence of the sensations which this most amazing spectacle created. More profound intellects than ours may, perhaps, be able to supply a satisfactory explanation of the mysteries of this singular tree ; but as to us, we altogether give it up." Botanists whose severe love of truth overcomes in most cases their poetical inclinations, have thrown considerable doubt upon this story, even though related by missionaries of a respectable character. It appears to be in some particulars considerably indebted to an ardent imagination, but it may, nevertheless, be true enough in its main facts. Divested of its apparent embellishments and exaggerations, the tree may be found after all to be only an exotic species of plane or sycamore, covered with immense patches of the written lichen, which —it is well known to botanists—occurs in greater

profusion and attains a larger size in tropical
than in temperate countries. Many exotic, and
one or two European lichens occur on living
leaves. These are principally developed on the
upper surface, sometimes only superficially con-
·nected with the leaves, which afford them a basis
of attachment and growth ; and at other times
originating like the fungi beneath the true cuticle,
forming a carbonaceous, beautifully-sculptured
crust and elegant fructification. The foliage of
the Thibetian wonder may, therefore, be indebted
for its singular markings to a species of Limboria;
and the characters on the bark and branches may
have been caused by an unknown Opegrapha.
In fact, the counterpart of these inscriptions has
been discovered by Hooker and Thomson in
Khasya, on the leaves of a species of Symplocos.

Let us glance at some of the peculiarities of the
lichens, and see if nature has not assigned them a
higher and more important commission in her
great household, than merely ornamenting old
walls and ruins, and covering trees with a shaggy
mantle.

The lichens have apparently no affinity with
the mosses. Their appearance is altogether different; and yet they approach the frondose division of the Hepaticæ or scale-mosses by the genus
Endocarpon, which consists of a greenish leathery

frond, and frequently flourishes upon constantly dripping rocks within the spray of waterfalls, a peculiarity contrary to the habit of lichens in general. The round fructification, deeply imbedded in the thallus, is similar to that of the Riccias or crystal-worts. This genus of lichens may therefore be fairly regarded as a connecting link with the scale-mosses. Indeed Fee placed it apart from the true lichens, in a section which he called Pseudo-Hepaticæ. With the fungi the lichens are most closely allied; the principal difference between many species on both sides being that the lichens possess a crust or thallus, while the fungi are destitute of it, and that the lichens grow on inorganic substances, and living structures, while fungi require dead and decaying organic substances as a matrix. There are other differences of course between the two orders, but they are of a microscopic character. Superficially the mutual resemblance is often perfect; and several Pezizas and Sphærias among the fungi have been constantly mistaken for lichens. To the algæ or sea-weed tribe the lichens are related on the one side by the *Lichinas*, which form blackish green cartilaginous fronds on rocks on the sea-shore, which are exposed and almost dry at low water, and which have received their name from their similarity to some of the lichen

family, among which indeed Acharius had placed
them ; and on the other hand by the Collemas, a soft
gelatinous family of lichens, found in damp shady
places, which alter their form and consistence
very much in drying, and assume very variable
shapes. Indeed the opinion has been recently
expressed by Continental botanists, that these
Collemas are only perfectly developed states
of plants, whose young, or imperfectly developed
forms have hitherto stood amongst the unicellular
or rudimentary algæ under distinct generic and
specific names. Some have gone even further,
and applied that observation to the whole tribe
of lichens ; asserting that the bud-like cells or
gonidia of every lichen, which have the power of
continuing to live and develop themselves even
when separated from the thalli which produced
them, may be referred to some species of low-
type alga. It may also be mentioned that Mr.
Sorby, while recently conducting researches among
the chromatological relations of the lower plants
and animals, has found a most interesting series of
connecting links between olive algæ and lichens
in the colouring matters which they yielded. All
these views countenance the opinion of Dr. Tucker-
man, the late distinguished American lichenologist,
who defined lichens to be " perennial aërial algæ,"
and regarded them as due to the transformation

of marine algæ, which would be the first vegetable inhabitants of our globe, and which, on the
emergence of the land from the water, would
adapt themselves, by undergoing various modifications, to their new element and circumstances.

Lichens, I have said, are exceedingly simple in
their construction. They are composed of two
parts, the nutritive and the reproductive system.
The nutritive portion is called the thallus, which,
in the typical plant, spreads equally on all sides
from the original point of development, in the
form of an increasing circle ; the circumference of
which is often healthy and vigorous while the
central parts are decayed or completely wanting.
It is composed of two distinct tissues. The central
or medullary portion is composed of spherical
cells, filled with a green matter, which seem to be
the active vegetating part of the lichen. These
cells called gonidia frequently accumulate in
masses, burst through the layer above them,
and appear in the form of a green, tenacious
powder on the surface of the plant ; while they
are capable, if detached from the parent, of continuing the powers of cell-development, multiplying by sub-division and spreading out into
filaments which form the nucleus of new lichens.
They perform in the economy of lichens an office
analogous to that of the gemmæ or buds of the

higher cryptogams, and of the bulbils, stolons, etc., of the flowering plants. The green matter of the cells or endochrome is resolvèd into zoospores as in the confervæ or fresh water algæ, and in the stalked spores and reproductive cells of fungi,—a circumstance which brings lichens into close re- lationship in an important point with algæ and fungi. The external or cortical layer called the hypha, on the other hand, is supposed by some botanists to serve the same purpose in the eco- nomy of the lichen as the bark does in that of the tree, viz., as a protection to the lower, living layer, of the dead cellules of which it actually consists. In some species this outer covering is smooth, and in others covered with small hollows or pits, and sprinkled over with powdery warts. The lower surface of the lichen is usually of a paler colour than the upper, and is covered with hair or fibres which serve to fix the plant.

A curious theory has recently been promulgated by Continental botanists regarding the parasitic nature of all lichens. They do not form a dis- tinct order of vegetation as we have been in the habit of regarding them, but are supposed to be produced by a combination of fungi and those confervoid algæ which are universally distributed on bark, wood, rocks, and mosses, attaining their greatest development in moist and shady places.

The connexion between the two layers of the lichen, alluded to in the preceding paragraph, is not the result of simultaneous individual growth in one organism, but is due to parasitism. The colourless filamentous tissue of the corticolous layer or hypha is supposed to be a kind of spawn or mycelium belonging to some obscure fungus living parasitically upon the coloured cellular portion or gonidia, whose resemblance to certain unicellular or rudimentary algæ has been already observed. Schwendener, who first suggested this startling idea, remarks that he saw the threads of the hypha of lichens penetrating the fronds of different primitive algæ, such as nostoc, growing beside them, encompassing the filaments of the alga with a net-work which swelled and extended itself at the points of contact, and thus burst the filaments of the alga enclosed into small fragments which became transformed into gonidia. "As the result of my researches," he says, "all these lichen-growths are not single plants, not individuals in the ordinary sense of the word ; they are rather colonies, which consist of hundreds and thousands of individuals, of which, however, only one plays the master, whilst the rest, in perpetual captivity, prepare the nutriment for themselves and their master. This master is a fungus of the class of Ascomycetes, a parasite

which is accustomed to live upon the work of
others ; its slaves are green algæ which it has
sought out, or indeed caught hold of, and compelled into its service.   It surrounds them, as a
spider its prey, with a fibrous net of narrow meshes
which is gradually converted into an impenetrable
covering ; but whilst the spider sucks its prey and
leaves it lying dead, the fungus incites the algæ
formed in its net to more rapid activity, nay, to
more vigorous increase."   Other lichenists, such as
M. Bornet, have adopted this view as the only one
capable of explaining satisfactorily all that has
hitherto been observed regarding the nature of
the thallus of lichens and their fructification.
But plausible as the hypothesis looks, it seems
to me to be destitute of foundation.   That the
gonidia of lichens are analogous to or even
identical with those of algæ, and that lichens and
fungi have between them no absolute line of
demarcation, is admitted on all hands ; but that
the relation of the hypha to the gonidia is that
of the spawn or mycellium of a species of mould-
fungus to the alga, and such as necessarily to imply
the idea of the one being parasitic upon the other,
may be more than doubted.   It is well known
that parasitic fungi destroy the living organisms
upon which they fasten ; and if this assumed
parasitic fungus does not destroy its assumed

algal host, but on the contrary excites it to more active growth and more enlarged production of tissue, then it is clear that it cannot be a fungus, but what we have believed it to be, the vegetative tissue of a veritable lichen. Mr. Berkeley remarks that he had seen the gonidia of a Parmelia, a species of foliaceous lichen, originating from hyphæ within the cells of some drift wood from the Arctic regions without the presence of any alga. And were any other argument needed to refute the hypothesis it might be found in the symmetrical and simultaneous growth of lichens, which is utterly contrary to what would take place were one portion parasitic upon the other, or were the lichens compound organisms made up by the union of fungi and algæ. Whatever may be said regarding the parasitism of the gelatinous Collemas,—which might with justice be excluded from the lichens altogether,—I do not know any alga which could be transformed by the influence of any fungus into the highly organized texture of the shrubby and foliaceous lichens such as Usnea, Cladonia, Cetraria, and the higher Parmelias.

Nature has bestowed upon the lichens a peculiar mode of reproduction which appears quite different from that of the higher orders of the vegetable kingdom ; and yet they are propa-

gated with as unerring certainty and as great
rapidity as the most prolific family of flowers.
Every one who has an attentive eye must have
often noticed the curious round disks or shields,
usually of a different colour from the rest of the
plant, with which their surface is often studded.
These are called apothecia, and correspond with
the flowers of the higher plants ; for in them are
lodged the seeds or germs by which the lichens
are perpetuated. When examined under the
microscope they are found to consist of a number
of delicate flask-shaped cells, called thecæ, containing 4, 8, 12, or 16 sporidia, that is, cells of an
oval form, with spores or seeds in their interior.
The mode in which these spores are ejected
affords as wonderful a proof of design as was seen
in the case of the ferns and mosses. It is
principally in moist or rainy weather that this
curious process is performed. When the entire
apothecium or shield is wetted, the layer bearing
the thecæ or seed-vessels becomes bulged out
above, whence arises a pressure on them, which
ultimately bursts them at the summit, and causes
the expulsion of their contents. Few things can
exceed in beauty, as microscopical objects, the
sporidia of many of the lichens. Some are bright
scarlet, others deep blue, and others green, olive,
golden yellow, or brown.

Besides these true organs of fructification, the lichens are furnished with other parts which possess the power of reproduction. A great many species, placed in unfavourable circumstances, seldom or never produce proper receptacles of seed; but this is no obstacle to their propagation, as their whole surface is covered with collections of free powdery grains, which germinate into new plants wherever they are carried by the winds. There are also present on some lichens spongy excrescences which resemble minute trees; and one peculiar genus is possessed of tubercles which occur on the back part of the frond, and are lodged in little cups which appear empty as soon as they have fallen out. The recent researches of the French lichenists, Tulasne and Itzigsohn, have discovered another kind of fructification which is very common and exceedingly interesting. This consists of minute, blackish, elevated, somewhat gelatinous points called spermogonia, occurring on various parts of the upper surface of the thallus. These resemble, in external appearance, the tubercular apothecia of the Lecideas; but their internal structure, as shown in Fig. 8, is quite different. They consist of little cavities or utricles opening on the summit by a tiny orifice, and filled with a thin transparent mucilage, in which is contained a number of

linear filaments of extreme tenuity, and somewhat curved, which vibrate slowly in every direction. These curious bodies are supposed to be
analogous to the spermatozoids produced in the
antheridia of the algæ and mosses, and which
seem to perform an essential part in the reproduction of all cryptogamic plants. By the discovery of these curious bodies in lichens and
fungi, the law of the duality of the organs of
reproduction, which was so long supposed to
be the exclusive privilege of the flowering plants,
is now found to be without an exception in the
vegetable kingdom. There is no really agamous
plant, no plant without sex. The name of cryptogamia, given to the flowerless plants under
the idea that their mode of reproduction was
altogether anomalous, is now a misnomer. Assi-
duous observation, and the perfection with which
microscopes are now constructed, have enabled
modern botanists to determine that in all plants, no
matter to what group they belong, flowering or
flowerless, there exist two distinct orders of reproductive organs, the relative value of which
may be compared to that of the two sexes in
animals. The segregation of certain parts from
the general organism, in order to fulfil more
effectually the purpose of multiplying the individuals of a species, is traceable in the humblest

and minutest lichens. New modes of reproduction are superadded to the primary one; and all these kinds of fructification are sometimes found on one plant at the same time, each of them being . capable, under certain conditions, of producing perfect individuals similar to the parent plant. It must not be supposed, however, that they all exercise their functions at one and the same

FIG. 8.—UMBILICARIA POLYMORPHA.

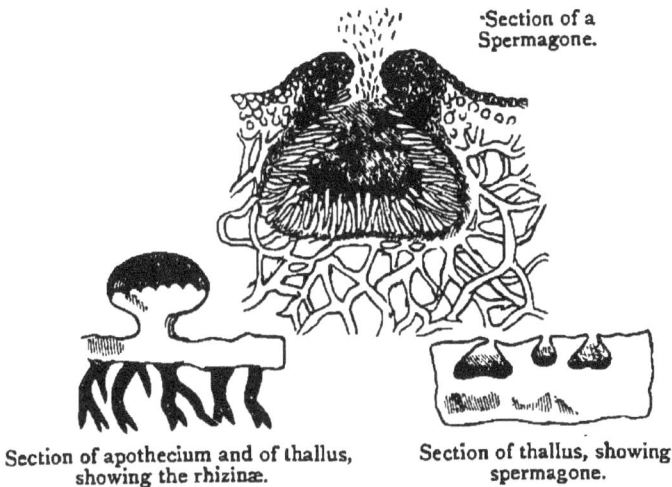

Section of a Spermagone.

Section of apothecium and of thallus, showing the rhizinæ.

Section of thallus, showing spermagone.

time—for nature is never prodigally wasteful of her resources; but where situation, temperature, or other conditions interrupt propagation by one mode, another is developed more exuberantly than usual to supply its place. If there be not conditions to produce perfect apothecia, there will be soridia, pulvinuli, or cyphellæ instead; and just as the chances of failure are great, so are

H

the modes of reproduction increased. And what
an admirable provision is this for the preservation
of plants, which would otherwise be speedily exter-
minated, exposed as they are to the contingencies of being successively scorched, drenched, and
frozen on the same naked and barren rocks! And
how greatly does it exalt these humble plants in
our estimation! Gifted with such powers of reproduction as these, we can view the smallest
lichen, not as a· single phyton, not as a single
frond, but as the aggregate of, it may be, thousands of these, view it occupying as much space,
and exercising as great an influence in the economy of nature as the largest forest tree, and
rivalling it even in longevity.

Lichens are very slow-growing plants. They
spring up somewhat rapidly during the first year
or two, as is evinced by the luxurious growth
which they form over young fruit-trees and
espaliers in gardens; but after a circular frond is
formed, they subside into a dormant state, in
which they remain unaltered for many years.
Mr. Berkeley says that he watched individuals
for twenty-five years, which are now much in the
same condition as they were when they first
attracted his notice. Some of the grey rosettes
of Parmelia which occur on walls and rocks, not
unfrequently attaining a circumference of many

feet, must be very aged, judging by this standard. The foliaceous and shrubby species are the most fugacious, though even these have great powers of longevity. We have no data from which to ascertain the age of tartareous species, which adhere almost inseparably to stones. Some of them are probably as old as any living organisms that exist on the earth. The geographical lichen, which often spreads over the whole rocky summit of a mountain in one continuous patch, many separate individuals being absorbed in one, must date from very remote periods. I have gathered it in this form on the summit of Schiehallion, on smooth quartz rocks which exhibit here and there the glassy polish and deep striæ or flutings peculiar to glaciated surfaces, as distinct and unchanged by atmospheric disintegration as though the glacier, which had left these unmistakable traces behind it, had only yesterday passed over them. And if these ice-marks can be accepted as an indication of the age of the lichen—the first and sole organic covering of the rock, be it remembered—then in all probability it was in existence during the last great changes of the globe which preceded the introduction of the human race. I do not press this point, however, for such a method of computation may be objected to; but I think that it is at least as reasonable to believe, that some lichens date their origin as far back as the glacial epoch, as to believe, that there are trees now in existence that were contemporaries of the first generations of men. There are numerous destructive and obstructive causes, fatal to the longevity of trees, which either do not operate at all, or only to a very limited extent in the economy of lichens ; and, indeed, these dry, sapless, dormant plants appear to me to possess the power of living for ever, without exhibiting any symptoms of decay, unless from accidental or extraneous causes.

We find in the shape and substance of the lichens an explanation of their stability and permanence. They are specially constructed both as regards form and composition for maintaining a low kind of vitality for indefinite periods, and for enduring in the midst of the most unfavourable circumstances. As a rule they are more or less spherical, and this is the form of greatest security, because it possesses the property of greatest symmetry or of equality in the relation of many points to one point and to one another. When growing they do not break out into straight lines which represent points separating from the general control and actuated by a single force in one direction, and are therefore unstable and insecure ; but they spread by ripples of circular growth, so that their stability is maintained at every point and stage of growth by their spherical form, which gives a maximum of contents with a minimum of exposure. We see in lichens as the form of their whole life, what in higher plants we observe only at certain stages of growth, and generally late in life, when they have exercised their active functions and returned to repose in the leaf, the blossom, the seed, the bud. And having thus a fuller and larger attainment of the spherical in form than other plants, we infer that they have a slower growth, a greater amount of stability and repose, and consequently a higher longevity. Nature, in the foliaceous webs of the lichens, works in her warp and woof as Penelope wrought at her loom, by fits and starts. She unrolls in a season of drought when no growth is made for weeks together, and the lichen seems withered and dead, what she had accomplished during a moist season, when the lichen was stimulated to new growth and exhibited the fresh green hue and the soft mobile tissue of active life. But not only is the form of the lichen thus suited to long suspensions of growth ; its substance also favours the retention of life in unfavourable circumstances, when endurance is the only quality which the plant can display. Lichens as a class are very largely composed of starch, which appears in cereal grains, in the tubers of potatoes, in fruits, in the wood of forest trees, and in all the parts of plants in which active life is suspended, owing to completion of growth or function, or unfavourable circumstances, such as those of winter or a dry season, and the organism returns to a state of rest. It is nature's admirable provision for keeping the fire of life in existence, until such times as it can start forth afresh in more favourable circumstances, by covering it over with its own ashes, as the thrifty housewife does with the embers on her hearth. When rain comes, and the lichen awakes from its dormant state, the starch which favoured its hybernation is utilized and transformed into materials of growth ; just as in our own muscles and liver, where starch, which was long supposed to be a peculiarly vegetable product, has been recently found in the form of glycogen as a normal constituent, is consumed in muscular and digestive action, and forms part of the fuel with which our muscular and hepatic engines are fed. In our own bodies, as in the humble economy of the lichen, starch contributes alternately to the repose and activity of life.

In their geographical distribution, lichens to a certain extent obey the same laws to which the higher orders of vegetation are subject, being influenced by temperature, altitude, and the geological character of the rocks upon which they are produced ; and thus several species and even genera are necessarily rare and confined to particular localities. It may, however, be said of them in general that they are cosmopolitan, universally distributed over the surface of the globe, and capable of existing in almost every situation, from the calcined plains of Africa to the snow-mantled pinnacles of Spitzbergen. Placed almost at the lowest scale of organization, they often require nothing more for their conservation, than the moisture of the atmosphere precipitated on naked masses of rock ; and their simple form and structure enable them to resist an amount alike of heat and cold, sufficient to destroy all vitality in more perfectly organized plants. In the Arctic regions—those outer boundaries of the earth, where eternal winter presides—these humble plants constitute by far the largest proportion of the flora, and by their prodigious development, and their wide social distribution, give as marked and peculiar a character to the scenery, as the palms and tree-ferns impart to the landscapes of the tropics. In the southern hemisphere also, lichens almost extend to the pole. They mark the extreme limit at which land vegetation has been found ; one shrubby species, with large, deep, chestnut-coloured fructification, called *Usnea fas-*

ciata, having been observed by Lieutenant
Kendal on Deception Island, the Ultima Thule
of the Antarctic regions. 'There was nothing,'
he says, in his interesting account of his visit to
that island, 'in the shape of vegetation except a
small kind of lichen, whose efforts seemed almost
ineffectual to maintain its existence among the
scanty soil afforded by the penguin's dung.' Dr.
Hooker also mentions that on this island he
found a few species of the beautiful pale green
Usnea melaxantha, looking like a miniature
shrubbery on the barren rocks ; on another island,
a few filmy specks of Lecanora and Lecidea, and
five peculiar mosses ; but that on Franklin
Island, and the islands nearer the Southern Pole,
he could not perceive the smallest trace of vegetation, not even a solitary lichen or piece of sea-
weed clinging to the rocks. Surrounded by huge
precipices of black lava, which seemed to fringe
them with mourning, and consisting entirely of
jagged rocks, formed of a kind of iron sponge
whose every pore has been filled with fire, covered
only with a little red soil, scorched and sterile, or
glittering snow-white patches of fragile shells and
coral, ground to dust by the fury of the waves,—
these remote islands exhibited an aspect so savage
and repulsive, so utterly lonely and lifeless, as to
impress with horror the stoutest heart.

Strange it seems that, while such extreme destitution, such sublime barrenness, prevails in these southern lands, in the Arctic regions, on the contrary, no spot has yet been discovered wholly destitute of vegetable life. The difference appears to arise more from the want of warmth in summer, than from the greater degree of cold in winter. The portion of heat imbibed by the soil, during the short summer of the Arctic regions, is prevented from escaping by the covering of snow which falls in the beginning of winter; and thus the temperature necessary for the scanty vegetation is preserved, till the return of the sun at once converts the Arctic winter into tropical summer, without the intervention of spring. Whereas in the Antarctic regions, the soil, owing to the much smaller quantity of snow that lies on it, is exposed to great alterations of temperature, which no vegetation, however simple and tenacious of life, can long successfully resist.

In the deserts of Asia and Africa, and on the coast of Peru, botanists have wandered for many leagues, without finding any other trace of vegetation than a species of grey or yellow lichen, growing on the blanched and mouldering bones of animals that had perished by the way. In tropical countries, where there is not too much moisture and shade, the trees are shaggy with lichens ; and some of the most magnificent species, both as regards size and colour, have been gathered in the Cinchona forests which clothe the lower slopes of the Andes, and in the warmer and more densely-wooded parts of Australia and New Zealand. The thick impervious forests of Brazil, however, are said to be almost destitute of them ; their places on the trunks and boughs of the trees being occupied by endless varieties of ferns, tillandsias, orchids, and other epiphytic plants, which seem to hold a floral revel ; the amazing luxuriance of higher vegetable life effectually keeping down and banishing plants of a simpler structure, and of a more sluggish and feeble nature. On the loftiest mountains of the globe they constitute the last remnants of vegetation, the last efforts of expiring nature which fringe the limits of eternal snow ; and long after the botanist has left behind him the last stunted Alpine flower, blooming like a lone star on a midnight sky, amid the loose crumbling stones of the *moraine ;* long after the last moss has ceased to deck the brown and lifeless ground with a scarce perceptible film of green, his eye, wearied by the universal desolation, rests with peculiar interest and pleasure on the hardy lichens, which clothe every rugged rock that lifts up its head through the avalanche, and which luxuriate amid the rack of the higher clouds and the howling of glacier winds. On the Alps of Switzerland the last lichens are to be found on the highest summits, attached to projecting rocks, exposed to the scorching heats of summer and the fierce blasts of winter; and from forty to forty-five kinds have been found in spots, surrounded by extensive masses of snow, between 10,000 and 14,780 feet above the level of the sea. It is interesting to know, that the only plant found by

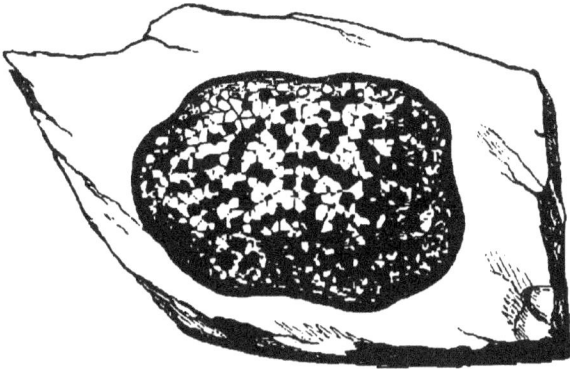

FIG. 9.— LECIDEA GEOGRAPHICA.

Agassiz near the top of Mont Blanc, was the *Lecidea geographica* (Fig. 9), a very beautiful lichen, which covers the exposed rocks on the sides and summits of all our British hills, with its bright-green map-like patches. This species was also gathered by Dr. Hooker at an elevation of 19,000 feet on the Himalayas, and occupied the last outpost of vegetation which gladdened the

eyes of the illustrious Humboldt, when standing within a few hundred feet of the summit of
Chimborazo, the highest peak of the Andes.
Strange it must have seemed to this enterprising
traveller to stand on that elevated spot, and to
see around and beneath him an epitome, as it
were, of what takes place on a grander scale over
the whole globe—a condensed picture of all the
climates of the earth from the tropics to the poles,
with all their different zones or belts of vegetation.
Above towered the inaccessible summit in its
everlasting shroud of stainless snow, boldly relieved against the deep cloudless blue of the
tropical sky; around him the bare and rugged
trachytic rocks, enamelled with the primrose-
coloured crust of this beautiful lichen, a few pale
tufts of moss, or a solitary flower drooping here
and there its frail head from a crevice ; immediately beneath him the green grass-clad slopes,
variegated with rainbow-coloured flowers and
stunted willow-like shrubs ; and far down in the
valleys at the base, a glowing gorgeous world of
tropical luxuriance—palms and bananas and bam-
boos, dimly revealed through the seething, sweltering vapours which perpetually surrounded
them.

The *Lecidea geographica* affords, I may mention,
the most remarkable example of the almost

universal diffusion of lichens, being the most
Arctic, Antarctic, and Alpine lichen in the world
—facing the savage cliffs of Melville Island in
the extreme north, clinging to the volcanic rocks
of Deception Island in the extreme south, and
scaling the towering peak of Kinchin-junga, the
most elevated spot on the surface of the earth. A
catholic beauty, it is to be found in every zone of
altitude and latitude—'a pilgrim bold in Nature's
care.'

On the British mountains we find lichens in
great abundance and luxuriance, in spots which
favour their growth by the humidity continually
precipitated from the atmosphere. Most of the
species found sparingly scattered at the highest
elevations are identical with those found in the
greatest profusion covering immense areas on the
plains of Lapland, and on the level of the sea-
shore in the Arctic regions ; the isotherms or lines
of equal temperature passing through these points.
Similar species are also found all over the world
below the level of perpetual snow, which on the
Alps is 7000 feet, and on the Andes and Himalayas about 15,000 feet. It is somewhat remarkable that Alpine lichens generally are more or less
of a brown or black colour. This peculiarity seems
to be owing to the presence of usnine or usnic
acid, which in a pure state is of a green colour,

as in the lichens which grow in shady forests,
but which becomes oxidized, and changes to
every shade of brown and black, when exposed
to the powerful agencies of light and heat on
the bleak barren rocks on the mountain side and
summit. These gloomy lichens, associated as
they almost always are with the dusky tufts of
that singular genus of mosses the Andræas, give a
very marked and peculiar character to many of
the Highland mountains, especially to the summit
of Ben Nevis, where they creep, in the utmost
profusion, over the fragments of abraded rocks
which strew the ground on every side, otherwise
bare and leafless, as was the world on the first
morning of creation, and ·reminding one of the
ruins of some stupendous castle, or the battle-field
of the Titans. Some of the Alpine lichens, however, are remarkable for the vividness and brilliancy
of their colours. The mountain cup-moss, with its
light-green stalk clothed and filligreed with scales,
and emerald cup studded round with rich scarlet
knobs, presents no unapt resemblance to a double
red daisy. It grows in large clusters on the bare
storm-scalped ridges, and forms a kind of miniature flower-garden in the Alpine wilderness. The
loveliest, however, of all the mountain lichens is
the *Solorina crocea*, which spreads over the loose
mould in the clefts of rocks, and on the fragments

of comminuted schist on the summits of the
highest Highland mountains, forming patches of
the most beautiful and vivid green, varied, when
the under-side of the lobes is curled up, by
reticulations of a very rich orange-saffron colour.
This species is not found at a lower elevation
than 4000 feet ; hence it is unknown in England,
Ireland, and Wales, whose highest mountains fall
considerably short of this altitude. I have
gathered it on Cairngorm, Ben Macdhui, and Ben
Lawers. In this last locality, which is well
known to botanists as exhibiting a perfect garden
of rare and beautiful Alpine plants, it grows in
greater abundance, I believe, than in any other
spot in the Highlands. It occupies the whole
ridge of rugged and splintered rocks, marked by
the tear and wear of elemental wars during
countless ages, which runs along the summit of
the hill. The surface of these rocks is covered
with masses of sharp abraded stones, interspersed
with meagre tufts of grass and moss ; and among
these the saffron Solorina luxuriates in large
patches. With what delight have I seen this
beautiful lichen, beaming out on me from its
dreary and desolate home, in the blustering days
of early April, when the snow was falling thick
around, and the howling wind sweeping by with
unobstructed keenness ! With fingers almost

benumbed with the cold, I have picked it up to
admire its beauty—a beauty, such is the arrogant
idea which man entertains of his own importance
in the world—which seems utterly thrown away
in a spot where human foot and human eye rarely
if ever rest. How often among those wildly
desolate and pathless solitudes, where one may
wander for whole days without catching a glimpse
of a single living thing, save perhaps some raven
on its way to its nest, leaving behind it the blue
sky without speck or cloud, or a ptarmigan
scarcely distinguishable from the grey rocks
around, winging its slow wheeling flight to the
neighbouring hills, and uttering its soft clucking
cry ; or when standing on some lofty storm-riven
summit, cut off from the rest of creation, by the
howling mists that come writhing up from the
dark abysses on every side, and as lone as a
shipwrecked mariner on some desolate island in
the sea, thousands of miles from any shore ; how
often amid such dreary scenes does a little wild-
flower, or even lowlier fern or lichen, arrest the
weary eye by its simple and mute appeal, and
awaken thoughts and sympathies which are never
felt, or at least allowed their full sway, amid the
busy haunts of men. Like the little moss which
revived the spirits of the lonely and despairing
Park in the African desert, it carries us back to

the populous world we had well-nigh forgotten, reminds us of the enjoyments and affections of home, and more than all, raises our thoughts to the Maker of the great and the small, who placed it there to cheer by its presence the lonely wilderness, and whose wondrous skill and goodness its every petal, leaf, or frond declares in language, silent and unuttered, yet more eloquent than a thousand words.

Some species of lichens are confined to certain geological formations. *Stereocaulon paschale* is the first vegetation that appears on lava when it cools and hardens. On Etna, Vesuvius, Hecla, and the Canary Islands it occurs so plentifully as to whiten the volcanic rocks with its tufted coral-like masses. It is not confined to lava, however, for it is often found on different kinds of rocks in sub-alpine districts, although as a rule it prefers trappean, porphyritic, and other rocks of igneous origin. One lichen, the *Endocarpon sinopicum*, is found only on yellow hone schist, or on micaceous stones in walls that are strongly impregnated with iron ; its own rusty colour resembling the red stone called sinoper, and looking as though it had been stained with the oxide of iron in its matrix. Some lichens prefer granite, others flint and quartz, others calcareous rocks, others micaceous schist, and others sandstone and slate ; and they follow boulders or erratic blocks of these rocks into localities that are widely different from their native habitats. Lichens peculiar to the mountain summits may thus be found on lowland plains and even at the sea level. One of the most remarkable examples of the connexion between lichens and the lithological character of their basis of support, may be seen in the development of species of lichens, sub-alpine and arctic in character, and totally different from the surrounding lichen-flora, upon the huge boulders spread over the great North German Plain, which came originally from Norway and Sweden. Certain trees attract certain lichens, which rarely desert them for other trees. Many species are found most abundantly on pine-trees ; others on oak and ash, and others on the beech and birch ; their growth on these trees being determined by the facilities which the bark affords for attachment or nourishment. Some lichens are found on rocks on the sea-shore, and do not flourish inland beyond the reach of the salt spray wafted by the winds. Of these the most remarkable are the *Parmelia aquila,* and the *Ramalina scopulorum.* The former is easily distinguished by its tawny-brown, sun-burnt colour, and the very numerous and much divided narrow segments of the thallus. It covers rocks on almost all our sea-shores, especially in the west of Scotland, in the greatest abundance. The latter lichen is equally common, growing among Sea-pink and yellow Parmelias on rocks along the sea-shore, to which it gives a shaggy appearance by its long rigid greenish white tufts. On the Standing Stones of Stennis in Orkney it grows to a length of six or eight inches ; while the ancient sea-cliffs at Appin, near Oban, are fringed with immense masses of it nearly a foot in length, presenting, along with the varied and richly coloured flowering vegetation which adorns the ledges and crevices, a most picturesque sight. This Ramalina is found in all parts of the world, on the shores of New Zealand and the Antarctic Islands, as well as in the most northern regions, and is one of the most widely distributed lichens in the world. In temperate countries it occupies the same place on the sea-shore which the Orchil (*Roccella tinctoria*) does in tropical and southern zones. Another sea-coast lichen is the *Placodium canescens*, found abundantly on trees in England near the sea, and on walls and rocks in Scotland. It whitens the walls of Craigmillar Castle, and the rocks of Arthur's Seat and Salisbury Crags, near Edinburgh, and gives an appearance of being white-washed to the loose stone dike that runs up from the shore to Dunstaffnage Castle, near Oban. It is one of the loveliest of all our lichens, with its remarkably neat orbicular thallus of a snowy white colour, closely pressed to the stone, and plaited and lobed at the margin, contrasting beautifully with its central black apothecia, which, however, it rarely produces.

There are three splendid foliaceous lichens found in Britain, whose proper home is in the Tropics. They are members of a family—the Stictas— which attains the largest size, the greatest beauty, and the widest distribution in the forests of South America, the West Indies, New Zealand, and in the South Sea Islands. Of the three British species, I found a solitary specimen of one, the *Sticta crocata*, on a mossy rock in the 'Birks of Aberfeldy.' How it got there was a puzzling circumstance. It is distinguished by the bright yellow powder with which the tubercles or warts on the upper surface, and the little cavities among the down on the under surface, are covered, and which is more abundant, and of a richer golden colour in New Zealand specimens. This lichen occurs in three or four places in Britain very sparingly; but it has a very wide geographical range, being found on the mountains of New Zealand, in Jamaica, along the western slopes of the Andes, in the Sandwich Islands, down to the Straits of Magellan, and the Falkland Islands, in New Zealand,

Tasmania, and Australia. It grows to a magnificent size on the summit of Table Mountain, Cape of Good Hope. In Europe it is found, besides our own country, in Spain, Greece, Turkey, and Germany. Its northern limit at Inverary, lat. 56° N., singularly coincides with the latitude of the most southern habitat at Cape Horn. Another species of the triad of tropical lichens found in Britain is the *Sticta macrophylla.* It occurs nowhere else in Europe but in the south-west of Ireland on shady rocks beside the Turk Cascade, near Killarney, and on Cromagloun mountain. Before its discovery in these places in 1829, it had been known only as an inhabitant of the Mauritius, Madeira, and the Azores. It grows plentifully on rocks at Ribeiro Frio, Madeira, and fructifies abundantly there. It has also been found in South American forests on the bark of Cinchona trees. Its coriaceous thallus is imbricated, with flat blunt segments naked and smooth above, and clothed with brown fibres beneath. When fresh its colour is bright green, but it soon fades into a pale leathery brown, with a slight tinge of red on the edges of the segments. Its presence in Ireland, along with the remarkable Iberian flora which is found there, is a proof, as already remarked, of the western extension of the European and African continents, and the existence of the so-called continent of Atlantis. It is worthy of remark that, unlike all other plants, lichens are more widely distributed in proportion as they are higher in organization and more complex in structure; all the British species of the genera *Sticta, Usnea, Stereocaulon, Sphærophoron, Ramalina,* and *Cenomyce,* which exhibit the highest development of lichenose vegetation, being found in one or other of their numerous protean forms, under almost every condition of latitude, altitude, and climate, although it is only in the more southern regions, where the humidity and temperature are more uniform, that we find them in constant fructification. Many problems of great interest and difficulty are furnished by the geographical distribution of our native lichens. The restriction of the Stictas, for example, which extend over a wide range of the earth's surface, and have such powers of adaptation, to a few localities in this country, is a circumstance as singular as the parallel fact of the London Pride, which grows in the same locality as the Irish Sticta, and is confined exclusively to a few damp mountain climates, being nevertheless the most easily grown and propagated of all border plants in gardens, even in the very heart of our large cities. We cannot account for the rarity of certain lichens that are capable of very wide distribution, any more than we can account for

the fact that not a few of our common garden
vegetables which flourish far inland, without salt
in the soil or air, such as cabbage, beet, celery,
sea-kale, and asparagus, are natives of our own
sea-cliffs or salt marshes, never growing naturally
away from the influence of the saline air ; or for
this other fact that the same type of plants, and
even the same species, are often common to the
sea-shore and the summits of mountains, and confined to these localities, while they are nevertheless capable, as is proved in the case of the Thrift
or Sea-pink, of being cultivated in any soil or
situation.

The great object which nature intended to subserve by the universal diffusion of the lichens is
obviously that of preparing, by the disintegration
of hard and barren rocks, an organic soil in which
higher orders of vegetation may exist. Humble
and apparently insignificant as they are, it is to
them we owe the bright array of vegetable forms,
which contribute so largely to the beauty and
usefulness of the world we inhabit. They form
the first link in the chain of nature by which the
whole earth is covered with a robe of vegetation.
Their powdery crusts and little coloured cups,
drawing their nourishment in most part from the
surrounding atmosphere, extend themselves over
the naked and desolate rock, and form, by the

particles of sand into which they crumble its surface, and their own decaying tissues, a thin layer
of mould fit for the reception of the simplest
mosses. These in their turn, add their contribu-
tion of withered leaves, and increase the film of
soil ; others of a larger growth supplying their
places, and running themselves the same round of
growth and decay. Plants of a higher and yet
higher order gradually succeed each other, each
series binding together, and preparing for the
growth of its own species or of others, the loose
and incoherent mass of decaying tissues, sand,
and disintegrated soil which the previous occupants had left behind them. At length the
rock, once as bleak and desolate as though it had
been vomited from the depths of some vast vol-
cano, and on whose surface the smallest wild-
flower could not find a resting-place for its tiny
root, becomes a verdant meadow fit to support a
host of animals ; a rich garden of beautiful flowers
smiling in the sunshine ; or a wide expanse of
noble forest waving its billowy foliage in the passing breeze.

> ' Seeds to our eye invisible can find
>   On the rude rock the bed that fits their kind ;
>   There in the rugged soil they safely dwell,
>   Till showers and snows the subtle atoms swell,
>   And spread th' enduring foliage ; then we trace
>   The freckled flower upon the flinty base ;

These all increase, till in unnoticed years
The sterile rock as grey with age appears,
With coats of vegetation thinly spread,
Coat above coat, the living on the dead ;
These then dissolve to dust, and make a way
For bolder foliage nursed by their decay.'

Precisely the same effects are produced on the newly-formed coral islands of the Pacific. The winds or the waves waft thither the invisible spore of some lichen that may have had its birthplace on the rocks of the far-off Andes ; it finds a resting-place, and the few simple circumstances necessary for its development, in some sheltered nook where the dashing waves have ground the coral into glittering sand ; and through course of time it assumes a crust-like appearance, puts forth its organs of fructification, and sows around it a colony of similar individuals. These harbour the wind-wafted soil beneath their tiny leaves, and form, by their decomposition, a layer of mould to which new species are day after day adding their decaying tissues, until at last a sufficient soil has been deposited for the growth of the ferns, the bread-fruit, and cocoa-nut trees that have been wafted from the neighbouring-islands. And thus, through the agency of an all but invisible seed, developed into the lowliest form in which it is possible to conceive that life can be maintained, what was once a barren, solitary islet, where no sounds were heard but the ceaseless dashing of the waves against the snow-white reefs, or the shrill cries of some chance flock of sea-birds, that made it their temporary resting-place during their flight to some happier shore, has become a paradise of bloom and beauty where man takes up his abode, and finds every comfort that can minister to his simple tastes.

Even on the desolate rocks that jut out from the sides of lofty mountains, where the eagle or the condor builds its eyrie, these humble sappers and miners of the vegetable kingdom are busy, fulfilling the task appointed them in the great household of nature, and forming a layer of soil, which ever and anon, as soon as it is deposited, is carried down by the storm or the stream to fertilize the valleys at the base. Egypt is the gift of the Nile ; its rich alluvial soil has been brought down by the swollen waters of the sacred river from the mountains of Abyssinia, where it was formed, perhaps, by the agency of lichens and other Alpine plants, and precipitated in its present form over the barren sands of the Libyan desert. And who knows how much of the tropical fertility and luxuriance of the vast plains, which stretch onwards from the bases of the Andes and the Himalayas, may be owing to countless generations of lichens, working cease-

lessly far up on the inaccessible summits, amid
the icy rigour and sterility of an Arctic climate ?
This is not an extravagant supposition ; we see
every day the wonderful power of little things ;
and we find that the most gigantic results are
often dependent upon agencies minute and insignificant in their individual state, but irresistible in
an aggregate of countless myriads. It is a sublime
truth, and one worthy of universal acceptation, that
even in the smallest and most apparently useless
productions, the intelligent eye will often behold
some of the most splendid manifestations of God's
inscrutable wisdom and gracious goodness. The
bleak sterility of these lofty regions, where the
lichens perform their untiring operations under
circumstances where we should naturally suppose
life and organization alike impossible, is yet the
means of preserving the fertility of mighty territories which would otherwise become deserts.

The student of nature who has examined these
humble plants with sufficient attention must have
been often struck with wonder and admiration at
the peculiar fitness which they display for the
work to which they have been appointed, as the
pioneers or precursors of all other land vegetation.
What could be better adapted to withstand the fury
of the storms that beat upon their exposed places
of growth than the crustaceous, powdery, or leaf-

like expansions which they often assume, hard
and inseparable almost as a portion of the rock
itself? Then their capacity of extracting their
nourishment principally from the surrounding
atmosphere; the curious property which they
possess of continuing for years without undergoing any perceptible change; their strong persistent vitality by which they are able—when
scorched by the summer sunshine, deprived of all
their juices, and reduced to shapeless, hueless
masses, which crumble into powder under the
slightest touch of the hand or the foot—to revive
again when exposed to the genial influences of
the rain, assume their fairest forms and hues,
and develop their organs of fructification for the
dispersion of their kind; and lastly, the facility
with which they can replace portions of their
substance that have been torn away by storms,
broken by the tread of man, or eaten by animals;
all these qualities illustrate the wonderful adaptation, in their structure and habits, to the unfavourable circumstances in which they are often placed.
Furnished by such powers as these, wherever
they fasten their tiny fangs the process of disintegration commences; and though carried on
slowly and imperceptibly, though ages may elapse
before any apparent effects have been produced,
except the increase of individuals and the more

shaggy and picturesque appearance of the rocks, yet the object of that steady, ceaseless labour will one day be accomplished. And it is humiliating to the pride of man to find that the noble piles of architecture built by him as if for eternity, though apparently as solid as the rock out of which each individual stone had been hewn, and as hard as the famous Roman cement which had resisted the utmost efforts of Goth and Vandal, must yield in the end to the slow but persevering assaults of the most diminutive and contemptible vegetables, and be brought back again by these apparently feeble agents to the bosom of nature, out of which he had reared them with such labour and skill. Here, indeed, we have an illustration of that comprehensive saying of Melanchthon, ' The humble ones are the giants of the battle ;' here we have sermons in stones, lessons taught us by the lifeless lichens of the permanence of nature, and the never-ceasing change and decadence attendant upon all the works and possessions of man.

The objects which lichens subserve when they are produced on rocks and ruins are thus sufficiently obvious ; but it is not so easy to determine their precise use when growing on trees. It has been asserted by some writers that so far from being beneficial, they are absolutely prejudicial to the welfare of the forests in which they abound. Such individuals, however, it is evident, totally misapprehend the nature of these plants, for they extract their nourishment principally from the medium with which they are surrounded, and not from the matrix on which they are developed, or to which they are attached. The fungi are the only plants that are produced from decay and corruption, and maintain their existence by exhausting the vital juices of other plants. That lichens are not injurious to the plants on which they grow, is clearly proved in the case of Peruvian bark; for the specimens which are covered with healthy lichens abound more in the peculiar medicinal principle, and realize a larger price, than those which are bare and destitute of lichens; while, on the other hand, the bark that is covered with the beautiful *Hypochnus rubro-cinctus* and other fungi is utterly worthless, as these deadly parasites decompose all the substances upon which they fasten by the absorption of their nutritive matter. There is hardly a tree in the whole world which, at some stage or other of its existence, has not been covered with lichens. I have frequently observed the trees of a whole Highland forest, covered from head to foot with a dense shaggy garment of these plants, and yet maintaining, during the natural term of their existence, a green and healthy appearance. The species that grow upon trees, it must be observed, are generally very different from those which grow upon stones. There is a considerable preponderance of foliaceous and filamentous over crustaceous forms, and these, owing to the looseness of their hold upon the bark, being generally attached only by small roots in their centre, or by a single knot at one of their extremities, do not close up the breathing pores of the tree, or prevent that free circulation of air which is necessary for the healthy performance of all its functions. Indeed, I am disposed to think that lichens are not only harmless, but greatly beneficial to trees; for those who have paid particular attention to pines which grow in open and elevated situations, must have often noticed that, not only is their bark thicker and more rugged on the side most exposed to the prevailing winds and rains, but also that it is more densely covered with shaggy lichens, so as to afford considerable warmth and protection. The colder the climate, and the farther north we proceed, the more densely clothed with this picturesque garment of nature's providing do we find the trees and shrubs, on the same principle, one would imagine, as the hyperborean animals are covered with thick furs. Indeed, so universally are lichens and mosses produced on the north side of trees, that the American backwoods-man, and the Norwegian woodcutter, whose faculties of observation have been keenly educated by nature herself, often employ them as a rude but safe compass to guide them through the intricacies and tangled labyrinths of the primeval forests.

Such are some of the most obvious purposes which these humble plants serve in the economy of nature ; let us now direct our attention to a few of the uses to which man has applied them. This is the only point of importance connected with them in the estimation of many, especially of those who gauge the works of the Almighty by a dry utilitarian law, and see no beauty or interest in any object, except in so far as they can find some real or manifest utility in its existence. Judged by this standard, and weighed in the balance with pounds, shillings, and pence, the lichens will not be found wanting. On account of the large quantity of starchy matter which they contain, they often considerably contribute to, and sometimes even entirely form, the diet of man and animals in those dreary inhospitable regions where the wintry rigour, or the scorching heat of the climate, forbid all other kinds of vegetation to grow. Every one is familiar with the fact that the reindeer-moss (*Cladonia rangiferina*, Fig. 10) forms altogether the food of that animal during the prolonged northern winters. This lichen grows sparingly in little tufts among the heather in this country, and sometimes whitens the sides and plateaus of the Highland hills, covering bare and verdureless places where the snow first falls in winter, and lingers longest in summer ; but it is in the vast sandy plains called by the Laplanders Flechten-tundra and Moos-tundra, as lichens or

FIG. 10.—CLADONIA RANGIFERINA.

mosses predominate, which border the Arctic ocean, that it flourishes in the greatest profusion and luxuriance. There it completely covers the ground with its snowy tufts, and occupies as con-spicuous a place in the economy of nature as the grass in warmer regions. Linnæus says that no plant flourishes so luxuriantly as this in the pine-forests of Lapland, the surface of the soil being completely carpeted with it for many miles in

K

extent; and that if by an accident the forests
are burnt to the ground, in a very short time the
lichens re-appear, and resume all their original
vigour. These plains, he adds, which strangers
would call an accursed land, are fertile pastures
to the Laplander, who, in possession of a tract of
such country, deems himself a prosperous man.
There vast herds of reindeer roam at will, enjoy-
ing themselves where the horse, the camel, and
the elephant would perish. The reindeer is the
life, hope, and wealth of the inhabitants of those
dreary and inclement regions. It draws their
burdens with all the patience of the ass, yields
its milk with all the docility of the cow, and
transports its owner from place to place over the
snowy and frozen plains, with all the fleetness of
an Arabian horse. Its flesh serves for food; its
tendons for strings to their bows, and its thick-
furred skin for comfortable garments and bed-
clothes to protect them from the rigours of an
Arctic climate. And this useful animal is exclusively dependent upon an humble lichen for
its support. What a deep interest therefore in-
vests this otherwise insignificant plant! That
vast numbers of families, living in pastoral simplicity in the cheerless and inhospitable Polar
regions, should depend for their subsistence upon
the uncultured and abundant supply of a plant so

low in the scale of organization as this, is surely a striking proof of the great importance of even the smallest and meanest objects in nature.

When in Norway several years ago I saw a herd of reindeer feeding upon the reindeer-moss on the summit of one of the Dovrefjeld Mountains. The lichen presented a different appearance from the variety which grows in this country. It formed immense consistent masses nearly a foot in depth, of a beautiful cream colour, and of wonderful elasticity, springing up when the foot sank into it up to the ankle. The individual plants were exceedingly beautiful, richly and intricately branched. There are three distinct varieties of it to be seen in Norway, one found in forests and called *C. sylvatica*, one on the lower moors and called *C. alpestris*, and the finest of all which inhabits the highest mountain ranges, viz., the *C. grandis.* Like all lichens and Alpine plants it becomes more luxuriant and lovely the higher its range. On the Dovrefjeld it formed one of the loveliest spectacles of the kind upon which my eye ever rested. I found that in many parts of Norway it is used as winter fodder for the cattle. At the end of September it is scraped by means of large iron rakes into heaps, whose position is marked by tall poles ; and when the roads are made accessible by the first fall of snow, they are carried down to the

farms on sledges.   The reindeer-moss is also used
by the Finlanders and Laps for stuffing pillows
and mattresses.   Occasionally too it forms an
ingredient of the 'famine-bread' composed of a
little oatmeal mixed with sawdust and pounded
lichens, which the inhabitants use when sore
pressed in times of scarcity.

When the ground is covered with hard and
frozen snow, so that the reindeer cannot obtain its

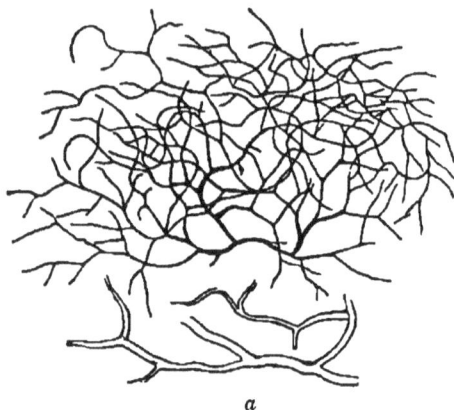

a

FIG. 11.—ALECTORIA JUBATA.
(*a*) Enlarged portion.

usual food, it finds a substitute in a very curious
lichen called rock-hair (*Alectoria jubata*, Fig. 11),
which covers with its beard-like tufts the trunk
of almost every tree.   In more severe winters, the
Laplanders cut down whole forests of the largest
trees, that their herds may be enabled to browse
at liberty upon the tufts which cover the higher
branches.   The vast dreary pine-forests of Lap-

land possess a character which is peculiarly their own, and are perhaps more singular in the eyes of the traveller than any other feature in the landscapes of that remote and desolate region. This character they owe to the immense number of lichens with which they abound. The ground, instead of grass, is carpeted with dense tufts of the reindeer moss, white as a shower of new-fallen snow ; while the trunks and branches of the trees are swollen far beyond their natural dimensions with huge, dusky, funereal bunches of the rock-hair, hanging down in masses, exhaling a damp earthy smell, like an old cellar, or stretching from tree to tree, in long festoons, waving with every breath of wind, and creating a perpetual melancholy twilight around. The Alectoria is found in great abundance in this country, especially in the pine-woods of the Highlands ; and is still employed in remote places as a stuffing for mattresses. In British Columbia, when all other food fails, the natives make shift with this lichen, which certainly does not look very nutritious. Commander Mayne describes it as one of the most important articles of food of the native Indians. They steep it in water until it is quite flaccid, and then, wrapping it up in grass and leaves to prevent its being burnt, they cook it between hot stones. They also boil it and press it into cakes three or four inches thick, which look like ginger-bread, but have a very earthy and rather bitter taste. 'Our companion,' says Commander Mayne, 'gave us this food, which the Indians call "Wheela," with milk. But two or three mouthfuls were all we cared to take.'

Another beard-like lichen (*Usnea florida*, Fig. 12), often growing along with the rock-hair, is gathered in great quantities in North America,

FIG. 12.—USNEA FLORIDA.

from the pine-forests, and stored up as winter fodder for cattle in inclement seasons. Goats, and especially deer, are fond of it ; and in winter, when other food is scarce, they hardly leave a vestige of it on the trees within their reach. The tortoises of the small rocky islands of the Galapagos Archipelago subsist almost entirely upon it. In this country it is one of the most picturesque ornaments of our pine forests. When fully developed it forms tufts nearly a foot in length. It is quite a miniature larch tree with root, stem, and most intricate branches and twigs. Its colour is pale sea-green ; and a central white thread or pith runs through the main stem, and lateral branches, on which, when cracked with age, the segments of cellular tissue are strung like beads on a necklace. A kind of farinaceous meal is plentifully sprinkled on the ultimate branches. Altogether it is one of the most beautiful and interesting of our native lichens. A reddish variety grows in such quantities on trees of Conyza arborea forming the alley near Napoleon Buonaparte's residence in St. Helena, that this hanging vegetation is the first thing that attracts the eye of the visitor.

But it is not to animals alone that lichens furnish a supply of food. Man himself is frequently directly indebted to them for subsistence. There are few, I presume, who are not acquainted with some particulars regarding the history and uses of that remarkable lichen, sold in chemists' shops under the name of *Cetraria Islandica,* or Ice-

FIG. 13. CETRARIA ISLANDICA. land moss (Fig. 13). Although in this country it is only used medicinally, as a

restorative diet in exhausting diseases, and during convalescence, for which it possesses an immemorial reputation, it forms one of the most important articles of food which the natives of Iceland possess. In fact, without it they would as certainly perish, as the favoured inhabitants of Britain without the more highly organized cereal plants, which, year after year, wave in all their golden beauty over the whole land, and are so strikingly suggestive of nature's bounty and munificence. What barley, rye, and oats are to the Indo-Caucasian races of Asia and Western Europe ; the olive, the grape, and the fig, to the inhabitants of the Mediterranean districts ; the date-palm to the Egyptian and Arabian ; rice to the Hindu ; and the tea-plant to the Chinese,—the Iceland moss is to the Laplanders, Icelanders, and Esquimaux.

In Scotland, the Iceland moss grows sparingly on the bare wind-swept sides and summits of the loftiest mountains, but in Iceland it is common over the whole surface of the country. It attains a large size on the lava of the western coast, and in the extensive desert tracts of Skaptar-fel-Syssel ; and numerous parties used to migrate to these places with all their household effects, during the summer months, in order to collect it, either for exportation to the Danish merchants, or for their own use as an article of common food. These excursions generally took place once every three years, for the lichen required that time to arrive at maturity, after the spots where it flourished had been cleared. Olafsen and Povelsen, in their interesting *Travels in Iceland,* observe that a family could collect four tons in a week during the season, with which, they say, they were better off than with one ton of wheat. We are also informed, in a report on this lichen, published several years ago by the Saxon Government, that the meal obtained from it, when mixed with wheat-flour, produces a greater quantity of bread, though perhaps of a less nutritious quality, than could be manufactured from the latter alone. Of gluten or nitrogenous flesh-forming material, it contains only one per cent.; but it contains no less than forty-seven per cent. of lichenine, which is a form of starch ; with three per cent. of sugar, and ten of gum and extractive. Its usefulness as an article of diet or of the Materia Medica must thus depend chiefly upon its lichenine or starch. The extremely bitter taste, however, by which it is characterized,—owing to a peculiar astringent principle in it called cetrarine which has been procured in a state of purity, in the form of a white powder like magnesia, by Herberger,—has always proved a great drawback to its adoption as an independent article of food, especially in this country. In Iceland and Lap-

land, however, the inhabitants remove this disagreeable quality by a very simple process. They
first chop it to pieces, and macerate it for several
days in water mixed with salt of tartar or quick-
lime, which it absorbs very freely ; it is then dried
and reduced to powder, and mixed with the flour
of the common knot-grass, made into a cake or
boiled, and eaten with rein-deer's milk, and eaten
with relish too, by these poor people, who confess,
with a most simple and affecting gratitude, that
' a bountiful Providence sends them bread out of
the very stones.' The powder is not unlike starch
in appearance, and possesses some of its properties,
for it swells in boiling water, and becomes, on
cooling, a fine jelly, which soon hardens into a
tough, transparent substance very pleasant to the
taste, especially when flavoured with sugar, milk,
a little white wine, or aromatics. It is frequently
used for making blanc-mange in this country, for
which purpose it is said to be equal, if not superior,
to Irish moss or the finest isinglass. The bitter
principle is often employed for brewing, and in the
composition of ship-biscuit, to prevent the attack
of worms. It also forms an ingredient of a well-
known form of cocoa called ' Iceland-moss Cocoa,'
as well as of a French confection known as 'Pâte
de Lichen.' It may be mentioned that notwithstanding its name, the Iceland-moss is not only

more plentiful, but more largely developed in
all its varied forms in Norway than in Iceland, and
it is in Norway that it is now almost exclusively
collected for the European market.

Those who have read the affecting account which
Franklin and Richardson give of their expedition
to Arctic America, must be familiar with the name
of the Tripe de Roche, which occurs on almost
every page, and is intimately associated with the
fearful sufferings which these brave men endured,

FIG. 14.—GYROPHORA CYLINDRICA.
(*a*) Enlarged portion.

a part of which only would have sufficed to unseat
the reason of most individuals. During their long
and terrible journey from the Coppermine River
to Fort Enterprise, one of the stations of the Hudson's Bay Company—a journey to which, I venture
to say, there are few parallels in the annals of
human hardship—in the almost total absence of
every other kind of salutary food, their lives were
supported by a bitter and nauseous lichen, to

which the name of Tripe de Roche (*Gyrophora,*
Fig. 14) has been given as if in mockery. I cannot resist the inclination to transcribe from this
melancholy narrative a single fragmentary passage
which will give some idea of the fearful condition
to which these heroic adventurers in the cause of
science were often reduced. I need not preface
it by any comment of mine ; it speaks for itself.
' Mr. Hood, who was now nearly exhausted, was
obliged to walk at a gentle pace in the rear, Dr.
Richardson kindly keeping beside him, whilst
Franklin led the foremost men, that he might
make them halt occasionally till the stragglers
came up. Credit, however, one of their most active
hunters, became lamentably weak, from the effects
of tripe de roche upon his constitution, and Vail-
lant, from the same cause, was getting daily more
emaciated. They only advanced six miles during
the day, and at night satisfied the cravings of hunger by a small quantity of tripe de roche, mixed
up with some scraps of roasted leather. Having
boiled and eaten the remains of their old shoes,
and every shred of leather which could be picked
up, they set forward at nine, like living skeletons,
advancing by inches, as it were, over bleak hills,
separated by equally barren valleys, which contained not the slightest trace of vegetation except
this eternal tripe de roche.' The dreadful un-

certainty, that for so many long years hung over
the fate of Franklin and his heroic comrades, was
at last dispelled by the discovery, during M'Clin-
tock's search, of a large cairn at Cape Victoria, in
King William Land, containing, among other
mournfully interesting relics, a journal of one of
the officers of the lost expedition, announcing the
intelligence of the certain death of its leader on
the 11th of June 1847. A short distance beyond
this fatal point, two human skeletons were found
in the bottom of an abandoned boat, with no food
beside them except some tea, chocolate and tripe
de roche, on which miserable and innutritious diet
they lingered out their existence in these frightful
solitudes, till death mercifully put an end to their
sufferings.

The tripe de roche consists of various species
of Gyrophora—black, leather-like lichens, studded
with small black points like coiled wire buttons,
and attached by an umbilical root, or by short
strong fibres to rocks on the mountains. Some
of them bear no unapt resemblance to a piece of
shagreen ; while others appear corroded, like a
fragment of burnt skin, as if the rock on which
they grew had been subjected to the action of
fire. They are found in cold exposed situations
on Alpine rocks of granite or micaceous schist, in
almost all parts of the world—on the Himalayas

and Andes as well as the British mountains. But it is in the Arctic regions alone that they luxuriate, covering the surface of every rock, to the level of the sea-shore, with a gloomy Plutonian vegetation, that seems like the charred cinders and shrivelled remains of former verdure and beauty. I gathered magnificent specimens of two members of the family *Umbilicaria pustulata,* and *U. spodochroa,* on rocks a little to the south of Christiansand, and on the rocks below the fortress in the harbour of Bergen in Norway. They grew in these two places in the utmost profusion, and to an enormous size; whereas on the heights of the Dovrefjeld, where I expected to see them still larger and more abundant, I found only a few dwarfed individuals. They seem to reach the maximum of their development in the extreme south and at the sea-level, decreasing in size and number as we proceed northward and ascend the high mountains. We are accustomed to regard the Umbilicaria pustulata as an Alpine lichen in this country, but I have never gathered it on mountains, and the only spot in which I have seen it attaining anything like the size of Norwegian specimens was on rocks of hypersthene around the shores of Loch Corruisk in the Isle of Skye. In that wild scene it grew to the size sometimes of half-a-foot in diameter ; and with its grey blisters rising above the dark charred-looking thallus, it imparted a weird aspect to the grim shores of that lake of Avernus. Though the Gyrophoras contain a considerable quantity of starch, they are exceedingly bitter and astringent, and produce intolerable griping pains when eaten. No one would have recourse to them for food except in a case of dire necessity. The Canadian hunters, who are often reduced to the last extremity, during their long and toilsome excursions in search of furs, through the desolate regions of Arctic America, often allay the pangs of hunger with this nauseous diet. And sometimes in my own wanderings among the almost unknown solitudes of the Scottish mountains, when my stock of provisions was exhausted, and a renewal was not to be expected, the nearest shepherd's sheiling being perhaps many miles distant, I have been compelled to satisfy my cravings by eating small portions of the tripe de roche, which I found blackening the dreary rocks around. In such situations, I have felt deeply how weak and helpless is man, when thrown forth from the social scenes and comforts of civilized life, left to his own unaided resources, and exposed to the merciless energies of physical nature, and how, without some ultimate trust in the Almighty source of his being, that being is but as a straw upon a whirlpool.

There are several other species of lichens, which have now and then, on rare occasions, been employed as articles of food. There is a greyish shaggy lichen abundant on pine-trees in the British woods, called *Evernia prunastri,* which is said in ancient times to have rivalled even the Iceland moss for its nutritious qualities. Forskoel says in reference to it in his *Flora Arabica,* ' I have heard a great deal about a Schoebean plant unknown to me, without a portion of which, mixed with its contents, no kind of bread is manufactured. Shiploads of it are regularly conveyed to Alexandria from the Grecian Archipelago. A handful of the lichen is inserted in water for two hours, which, when added to the dough, imparts to the bread a peculiar flavour, esteemed delicious by the Turks.' It is possessed of a mawkish insipid taste, and especially if produced on oaks is somewhat astringent, but not destitute of nutritious qualities.

There is a curious lichen found in some eastern countries called *Lecanora esculenta,* regarding which several strange facts have been related by travellers. Some authors are strongly of opinion, that the manna with which the Israelites were fed in the wilderness may be referred to this lichen. A pamphlet has been published upon the subject by Dr. Arthaud. Such a reference may be supposed by some to militate against the professedly miraculous character of the event. But this objection may be overruled by the consideration, that though the manna was miraculous, in so far as the manner of its conveyance to the Israelites, and the circumstances connected with its gathering, were concerned, it was not miraculous in its origin. The quails were conveyed to the Jewish camp by supernatural means, but they were not supernatural in themselves; and, in like manner, the manna was showered down by the direct agency of God, in the very place where, and at the very time that it was required; but it was not a miraculous substance; it was not specially created for that purpose. God is sparing of His miracles; and in all His direct interpositions on behalf of His people, we find that He makes use of objects and agencies already existing, causing these to fall in with His intentions, without originating new ones. If the manna was a vegetable product already existing, and not a special creation, there is more likelihood of its being a species of lichen, than any other vegetable matter which commentators have conjectured. The descriptions of Moses apply with greater accuracy to the *Lecanora esculenta*,

L

than to any other substance with which I am
acquainted; while the singular circumstances
connected with the history of·this lichen, as related from time to time by trustworthy witnesses,
renders the supposition of its identity with the
manna of the Israelites still more plausible.
Showers of this lichen have sometimes fallen
several inches thick, having been torn from the
spots where it grew, and transported by violent
gusts of wind. In 1829, during the war between
Persia and Russia, there was a great famine in
Oroomiah, south-west of the Caspian Sea. One
day, during a violent wind, the surface of the
country was covered with a lichen, which fell
from the sky in showers. The sheep immediately
attacked and devoured it eagerly, which suggested to the inhabitants the idea of reducing
it to flour, and making bread of it, which was
found to be palatable and nourishing. The
people affirmed that they had never seen this
lichen before or after that time. During the
siege of Herat, more recently, the papers mentioned a hail of manna which fell upon the city,
and served as food for the inhabitants. A rain
of manna occurred so late as April 1846, in the
government of Wilna, and formed a layer upon
the ground three or four inches in thickness. It
was of a greyish-white colour, rather hard,

irregular in form, inodorous and insipid. Pallas, the Russian naturalist, observed it on the arid mountains, and the calcareous portions of the Great Desert of Tartary. Mr. Eversham collected it on the steppes of the Kirghiz to the north of the Caspian Sea. It has been seen on the Altai range, in Anatolia, in South America, and recently in Algeria by Dr. Guyon. It occurs in irregular-shaped fragments, varying in size from a pin's-head to a pea or small nut; and when seen in its native sites, is apparently attached to no matrix whatever, and has no fecula in its composition.

In medicine, lichens were at one time very highly esteemed. In the days of Aldrovandus and Paracelsus, who added the study of alchemy and the occult sciences to that of plants, they were extensively employed in the preparation of sympathetic ointments, and in the various distillations connected with the search for the elixir-vitæ and the universal solvent and nostrum. Wonderful cures were ascribed to a particular application of them; and in the works of the botanists of the middle ages, we find long and elaborate observations upon the peculiar virtues of species developed upon the oak, the pine, and the beech. The common dog-lichen (*Peltidea canina*)—a species everywhere abundant on moist banks and turfy walls, and easily distinguished by its livid brown wrinkled leaves, and red, nail-like fructification—was formerly employed, at the suggestion of the celebrated Dr. Mead, as a cure for hydrophobia, hence its specific name ; but if it ever effected a cure, it may be pretty safely asserted that it was more by the aid of a strong imagination than by any inherent healing power in the plant itself. Another species of the same family (*Peltidea aphthosa*), with a remarkably vivid green thallus, growing by the side of mountain streams, was in high repute at one time as a powerful anthelmintic, and is still used by the Swedish peasants, when boiled with milk, as a cure for the aphthæ or thrush in children. When the primitive principle that 'like cures like' and the strange 'Doctrine of Signatures' as it was called, formed the basis of all medical treatment, several lichens were employed for the cure of diseases, on account of their fancied resemblance to the organs or parts of the body affected. Among such lichens the species in greatest favour was probably the lungwort (*Sticta pulmonaria*), which grows in immense shaggy masses on trees and rocks in subalpine woods. From the resemblance of its reticulated and lobed upper-surface, usually of a greyish-brown colour, to the human lungs, it was highly recommended as an infallible cure for all diseases of these delicate organs. The beautiful

cup-lichen, so abundant on dry moorlands under
the shade of the heather, was long a favourite rustic remedy in this country for coughs. Gerarde,
the old English herbalist, says : ' The powder of
this moss given unto small children, in any liquor
for certaine daies together, is a most certaine
remidy against that perilous maladie called the
chin-cough. Albeit the remidy doth require care,ᶜ
and is not to be adventured upon save under the
guidance of an experienced gude-wife.' On account of the intensely bitter principle contained
in greater or less degree in all lichens, many
species used to be employed in intermittent fevers
and agues, as substitutes for Peruvian bark, which
was then sold at a price so extravagant, as to be
utterly beyond the reach of the poorer classes.
For the same reason, they were often administered in the form of powders and decoctions, as
tonics to purify the blood and strengthen the
system. Their astringent qualities—depending,
I may remark, in a great measure upon the kind
of tree on which they were produced—were also
turned to advantage in the cure of hæmorrhages,
fluxes, and ruptures ; and Linnæus informs us
that the Laplanders fill up their snow-shoes with
one species, and apply it to the feet to relieve the
excoriations occasioned by long and fatiguing
journeys. During one period of medical history,

lichens formed the principal drugs in the pharma-
copœia, and were prescribed for almost all the ills
that flesh is heir to. Superstition had much to
do with their popularity in this respect. Their
strange shapes, their anomalous character, occupying, as it were, an intermediate position between
plants and minerals, between life and death ;
leading a perpetual mesmerized or suspended
existence ; the curious situations in which they
were found, growing on decaying wood or moist
earth, or on the bare rock in weird, lonely spots,
where fairies might sport and enchanters weave
their unhallowed spells ; they were naturally
enough supposed by a credulous and ignorant
people to be invested with magic qualities. As
the knowledge of plants became more generally
diffused, they lost much of their mystery, and
consequently of their power over disease ; and
now they have almost entirely disappeared from
medical practice. It must not be supposed, however, that they were thus summarily expelled
from the schools of medicine because they were
entirely destitute of healing qualities. Some of
them have been found, by chemical analysis, to
contain principles of great efficacy in certain complaints ; but as these principles varied in their
strength, according to the circumstances in which
the plants were produced, no dependence could be

placed upon the action of the doses administered.
It is obvious that the chemical qualities of cellular
plants, whose construction is so extremely simple,
must vary considerably in different individuals
and in different situations. The nature of the
matrix on which lichens grow, and of the medium
which surrounds them, must, to a great extent,
determine the presence in them of certain constituents which are extremely volatile, and dependent upon such conditions. The lichen that develops certain qualities when growing on the bark
of a tree, will not develop them to the same extent when growing on a rock ; and there will be
a similar, if not a greater difference between the
qualities of an individual produced in the shade of
a dark moist wood, and those of the same plant,
scorched by the sunshine and swept by the wind
on a bare exposed rock on the hill-side. It was
this variable chemical character, and the uncertain
medical results connected with it, that banished
the lichens from the druggists' shops. The discovery of new and more powerful drugs, obtained
from tropical plants stimulated by intense sunshine and highly organized soils, hastened their
exile, and effectually closed the door against their
return to favour ; while at the same time it greatly
diminished the list of native remedies, the products
of a cold, moist climate, and of poor and feeble

soils. The Iceland moss is the only species of lichen which has retained its place in modern pharmacy, as a tonic and febrifuge in ague ; but it is now principally employed, when added to soups and chocolate, as a palliative to consumption, and as an article of diet in the sick-room, and is being gradually superseded by the more nourishing productions of foreign countries.

It may seem strange that lichens should be employed in perfumery, considering that in themselves they are entirely destitute of odour, but such nevertheless is the case. The ancients appear to have been in the habit of using extensively a species of white filamentous lichen called Usnech, which grew upon trees in the islands of the East Indian Archipelago, St. Helena, and Madagascar, and exhaled, when moistened, an exceedingly agreeable fragrance, somewhat resembling musk or ambergris. This odour it may have derived from the spice trees on which it was produced. Among the Arabian physicians it was once in high repute when macerated in wine, as a cordial and soporific. So late as the seventeenth century, some of the filamentous lichens were sold in the shops of barbers and perfumers under the name of Usnea, and they formed the basis of a celebrated fragrant powder for the toilet, called Corps de Cypre gris or Cyprio, which is still manufactured on a large scale in Rome, and in some other cities of Italy. Their employment for this purpose, however, did not depend upon any peculiar inherent scent, for the species used are perfectly odourless, but upon their aptitude for absorbing and retaining, for almost any length of time, the fragrance communicated to them. Indeed, several of our tree-lichens possess in so remarkable a degree this curious property, that they are still employed in the manufacture of the most valuable and esteemed powder perfumes.

Various other substances useful in the arts and manufactures are yielded by the lichens. The late Lord Dundonald discovered a method of extracting from a species of white filamentous lichen (*Evernia prunastri*), very frequent upon pines and oaks, a kind of gum which was extensively used in Glasgow during the French war, as an efficient substitute for the expensive Gum Senegal, in calico-printing. When it was the absurd fashion to wear the hair whitened with powder, this same lichen was sometimes pulverized and employed, on account of its cheapness, instead of flower or starch. A species of yellow shrubby lichen, like brass wire (*Borrera flavicans*), found on apple and other fruit-trees in Devonshire, Sussex, and other parts of the south of England, used to be employed in Norway in poisoning wolves, which were at one

time a dreadful scourge in the country, ranging
the gloomy pine forests in immense herds, com-
mitting fearful havoc among the sheep-folds and
cattle-sheds, and when rendered desperate by
hunger, even attacking travelling parties and the
houses of the inhabitants. Chemists have detected
oxalic acid in several species of crustaceous lichens
growing on the bark of trees, and distinguished by
an intensely bitter taste ; and in one or two species
in such abundance, that 100 parts yielded 18 of
lime, combined with 29·4 of oxalic acid. The ox-
alate of lime bears the same relation to lichens as
carbonate of lime to the corals, and phosphate of
lime to the bony structure of the more highly
organized animals. On account of this circumstance, some of the crustaceous lichens are extensively employed in France in the manufacture of
oxalic acid ; and a small proportion of what is now
used in this country is derived from this source.
In London, various species of tree-lichens are sold
for the use of bird-stuffers, who line the inside of
their cases, and decorate the miniature trees upon
which the birds perch, with their shaggy leaves, so
as to give them a more picturesque and natural
appearance. The inhabitants of Smoland in
Sweden are said to scrape a peculiar species of
yellow crustaceous lichen from old pales, walls,
and rocks, and mix it with their tallow, to make

the beautiful golden candles which they burn on festival days. A wonderful race are these same Smolanders. They are so remarkably industrious and inventive, that they have given rise to a popular proverb in Sweden, ' Put a Smolander upon a roof, and he will get a livelihood.' ' This character,' says Frederika Bremer, in her charming work, *The Midnight Sun*, ' is strangely imprinted on the remote forest-regions of the country. The forest, which is the countryman's workshop, is his storehouse too. With the various lichens that grow upon the trees and rocks, he cures the diseases with which he is sometimes afflicted, dyes the articles of clothing which he wears, and poisons the noxious and dangerous animals which annoy him. The juniper and cranberry give him their berries, which he brews into drink ; he makes a conserve of them, and mixes their juices with his dry salt-meat, and is healthful and cheerful with these and with his labour, of which he makes a pleasure.'

If we wish to obtain a true idea of the value and importance of lichens in human economy, we must consider them in perhaps the most singular of their aspects, viz., as dye-stuffs and sources of colouring matter. Many of the tree-lichens, in a moist state, are very showy, yielding in water a coloured infusion corresponding to the hue of their

own leaves ; but strange to say, these are the least
valuable species to the dyer.   The lichens which
are richest in colorific principles are crustaceous
species growing on rocks, and utterly destitute of

FIG. 15.—ROCCELLA TINCTORIA.

colour in their natural state ; and it is one of the
most striking triumphs of chemistry as applied to
the arts and manufactures, that by its means some
of the finest shades of red, purple, and yellow are
extracted from such unlikely substances.   The
lichen popularly known as Orchil (Fig. 15) affords
a remarkable illustration of the extent to which
colorific principles are developed in these out-
wardly hueless plants.   It derives its generic name
Roccella from a Florentine family called Rucellai,
whose founder, for a long time a trader in the
Levant, discovered in the sixteenth century the
art of preparing a most valuable dye from it, by

the sale of which he realized in a short time a very large fortune. If, however, we are to believe Tournefort, the preparation of orchil was known to the ancient Greeks ; the purple of Amorgos, one of the Cyclades Islands, with which the celebrated tunics of the same name were dyed, being obtained from this lichen. Some authors are of opinion that it was the orchil, and not the little murex, a species of shell-fish found on the coast of Syria and Phœnicia, which supplied the famous Tyrian purple, the exclusive badge of imperial rank referred to in Ezekiel: 'Fine linen, with broidered work from Egypt, was that which thou spreadest forth to be thy sail ; blue and purple from the isles of Elishah was that which covered thee.' The frequent representation of the little shell-fish on the coins dug up among the ruins of Tyre must, however, be regarded as a sufficient refutation of this idea. The secret of the Rucellai was soon divulged, and the manufacture transferred to Holland, where a considerable trade in this lichen is still carried on. Orchil is found in small quantities on rocks by the sea-side in the extreme south of England, and in the Guernsey and Portland Isles. In warm climates, however, it occurs in profusion, especially on the volcanic rocks, and the sea-shores of the Canary and Cape de Verde Islands, in the numerous isles of the

Grecian Archipelago, and on the coasts of China and Peru. In the Indian collection of raw vegetable products exhibited in the Crystal Palace of 1851, several specimens of orchil from India, Ceylon, and Socotra were shown ; and an explanatory note appended to some from the bare, desolate Gibraltar of the Red Sea, the rock of Aden in Arabia, stated most suggestively—' Abundant, but unknown as an article of commerce.' It is probable that it occurs on the maritime rocks of all tropical countries in equal profusion.

In appearance this valuable lichen resembles a diminutive leafless shrub, forked, and subdivided into numerous roundish, irregular branches. It is tough and leathery in texture, of a whitish or blue grey colour, and covered with a mealy powder, or scattered warty excrescences. It is imported in the same state in which it is gathered from the volcanic rocks ; and those who prepare it for the use of the dyer grind it between stones, so as thoroughly to bruise but not to reduce it to powder, moistening it occasionally with ammonia mixed with quick-lime. By this process it acquires in a few days a purplish-red tinge, and is found to form a confused mass of violet-coloured threads. In this state it is employed to give the English broadcloths that peculiar lustre and purple tint, when viewed in a certain light, which are so much admired. When beaten to a pulp, and dried in little cubes about the size of dice, which have an azure colour with white spots, and an unpleasant odour, the orchil is called litmus. This substance contains, according to Gelis, three colouring principles : one soluble in ether, which is orange-red ; one soluble in alcohol, and one in water, both of which have a most beautiful purple tint, which they lose when excluded from the air, and regain when again exposed. On account of its exceeding delicacy, and the ease with which it may be applied, litmus is chemically used as a test of akalinity and acidity in the form of paper saturated with it, preserved in well-closed vessels, and secluded from the influence of light. This paper is turned red by an acid, and is restored to its original blue colour by an alkali. Orchil contains certain other substances, called orcine and erythrine, which are perfectly colourless, and contain no nitrogen ; but when exposed to the action of ammonia and common atmospheric air, they yield exquisitely beautiful colouring matters, which crystallize in regular flat quadrangular prisms, have a very sweet flavour, and of which nitrogen is an essential element.

In the Canary and Cape de Verde Islands, orchil was at one time the most important article of commerce ; the annual exportation

being valued at from £60,000 to £80,000 ; but
so great has been its consumption of late
years, that the best quality, which generally
sells for £200 a ton, and has in times of scarcity
been actually sold for the enormous sum of
£1000, or about 9s. a pound, has become exceedingly rare, and what is now commonly imported from other countries is worth little more
than £30 the ton.   By far the most valuable dye-
lichens known are various species of Roccella
growing on trees by the sea-side, at Zanzibar and
along the coast of Eastern Africa.   They come
to this country partly by Bombay, and partly
through Portugal and France, and have sup-
planted in the British market all other species of
lichens.   And yet in their native country it is
not known that they are capable of yielding
dyes or of being otherwise utilized in the domestic
arts.   In this country the sap of the *Roccella tinctoria* is of a deep yellow, staining the fingers when
gathered.   The colouring matter of all the species
separates itself and is easily obtained, if we rub the
lichen between the fingers when immersed in water,
which in consequence becomes milky.   When set-
tled a whitish powder is deposited, which contains
erythric acid.   The colouring matter is thus easily
expressed, because the surface of the thallus is not
covered by any cuticle, the cortical layer being

simply composed of short filaments placed close together and erect, so that their summits constitute the surface of the thallus. The summits of these filaments and the narrow interstices between them are sprinkled with a white powder. When saturated with water and rubbed by the hand, this powder comes away at once; and when liquefied by a solution of hypochlorite of lime, it instantaneously assumes a red colour, which is very fugitive. The same colour immediately appears if we apply this reactive to the surface of the thallus of the lichen. We are thus enabled to say what is the quantity of colourable matter which the different species and varieties of the genus contain. It is in fact a sort of immediate analysis. And it shows to us that the colourable matter is formed and excreted on the outside of the gonidial layer, there being but feeble traces of it towards the interior of the lichen or in the medullary layer; in this respect differing from other lichens, which contain the colourable matter underneath the gonidial layer and not upon it or in its exterior, so that it is necessary to cut their thallus and expose the medullary layer, whenever we wish to subject them to the test of the hypochlorite of lime.

In this country there are many species of lichens, growing in greater or less abundance, on the mountain rocks, which might be advantageously sub-

M

stituted for the rare and expensive foreign orchils.
Many of them have been known to the rural inhabitants from time immemorial. The parti-col-
oured and often exceedingly beautiful tartans of
the Highland clans, used to be dyed with the colouring matter derived from the common grey foliaceous lichens which so plentifully clothe almost
every tree and wall ; and many an old woman in
the remote parts of Scotland, skilled in the medicinal and dyeing properties of the various plants

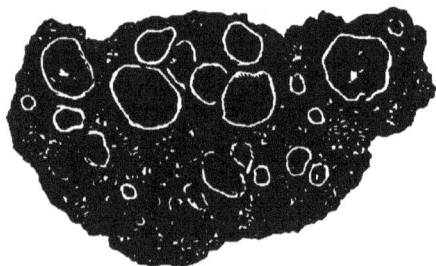

FIG. 16.—LECANORA TARTAREA.

that grow around her humble home, still prefers
the dyes she herself prepares, by simply boiling
in water heather twigs, birch leaves, roots of the
ruadh or yellow bed-straw, or the various species
of crotal or lichens, to logwood, madder, indigo,
copperas, or any other of the imported dyes of the
shops ; and the results she produces, by a skilful
combination of these simple substances, are really
astonishing ; many of the stuffs which have under-
gone her primitive dyeing process, being as brilliant and lasting in colour as those which have

been subjected to the various baths of the pro-
fessed dyer.

The most useful and best known of our native
dye-lichens is the rock-moss or cudbear, Fig. 16
(*Lecanora tartarea*), so called after·a Mr. Cuthbert
who first brought it into use. It grows in the form
of a tartareous granular crust, of a dirty-grey colour, spreading in indefinite patches over the surfaces of mountain rocks, and often enveloping the
stems and leaves of mosses and other small plants.
It varies in thickness from a scarce perceptible
film to a solid mass an inch in diameter, is covered
with large irregular shields of a pale flesh colour,
and may be easily identified, even without the aid
of its characteristic fructification, by a peculiar
pungent alkaline smell, which is very disagreeable,
especially when the plant is moistened. In Thors-
havn, in the Faroe Islands, it is so plentiful from
the sea-level up to the tops of the hills, that at a
distance it makes the stones appear as if covered
with lime. In the Highland districts, many an
industrious peasant used to earn a comfortable
living, by collecting this lichen with an iron hoop
from the moorland rocks, and sending it to the
Glasgow market. The value of this lichen in
Scotland is said at one time to have averaged £10
per ton. Hooker states that at Fort-Augustus, in
1807, a person could gain 14s. per week by gather-

ing it, estimating its market price at 3s. 4d. per
stone of 22lbs. It appears also to have been an
article of commerce in Derbyshire ; the price there
given to the collector, who could gather from 20
to 30 pounds per day, being 1d. per pound. This
source of remunerative employment in Britain has
now ceased, as the lichen is chiefly imported from
Norway and Sicily, where it occurs in greater profusion than with us, and is said to contain a larger
proportion of colouring matter. The dye produced
by the cudbear is quite equal to orchil, and is capable of being so modified as to give any tinge of
purple or crimson. It is never employed by itself
to give fast colours to cloth, but merely for the
purpose of improving the hues already imparted.
It is sold to the dyers in the form of a purple
powder. Schunk, in his analysis of this plant, discovered a colourless crystalline acid, called erythric acid, which is soluble in alkaline solutions,
and converted by them into orcine and carbonic
acid, and which, under exposure to the air, acquires first a red and at length a fine deep violet
tint.

A species closely connected with the cudbear,
and often growing together with it on the same
rock, is very extensively employed in the south of
France. This is the famous Perelle d'Auvergne
(*Lecanora parella*), which imparts those beautiful

and brilliant hues to French ribbons, which are so
much admired. The common yellow wall-lichen
(*Parmelia parietina*), so abundant everywhere,
yields a beautiful golden yellow crystallizable
colouring matter called chrysophanic acid, which is
identical with the yellow colouring matter of rhu-
barb ; and like orchil-litmus, it may be used as a
test for alkalies, as they invariably change its yellow colour into a vivid red tint. A beautiful and
valuable crimson pigment, occasionally employed
by artists, is the product of a dark-brown shrubby
lichen (*Cornicularia aculeata*), very common on the
hills ; while the common stone lichen (*Parmelia
saxatilis*), which forms grey rosettes on almost
every wall, rock, and tree, is still collected abundantly by the Scottish peasantry, under the name
of stane-raw, to dye woollen stuff of a dirty purple
or reddish-brown colour. On the low rocks, on
the summits of all the loftiest Highland hills, there
is a curious leafy lichen (*Parmelia fahlunensis*)
found abundantly, scorched apparently by the sun
into a black cinder. Of all lichens, this species,
judging from its outward colour and appearance,
would seem to be the least capable of yielding
colouring matter ; and yet, when treated in the
ordinary way, it yields a brilliant pink, cherry, or
claret colour, which in France has been applied to
so many useful purposes, that the lichen in con-

sequence has obtained the common name of ' Her-
pette des Tenturiers.' The same remarks may be
made regarding *Umbilicaria pustulata*, already
mentioned as an edible species, formerly largely
collected in Norway for the London orchil market, and known as the 'Pustulatous moss' of commerce ; and also regarding the *U. spodochroa*, the
largest and most plentiful of all the Norwegian Um-
bilicariæ, imported to this country for the manufacture of cudbear under the name of the ' Veluti-
nous moss.' But it is needless to enumerate all the
different species of lichens which have been, or are
still, employed in different parts of the world in
the production of colouring matter. This is the
characteristic quality, more or less, of the whole
tribe. The whole world may be said to be an
open field ; in every clime, in every soil, at almost
every elevation, and in all seasons, tinctorial species
grow, and even luxuriate. It is a matter of surprise in this age of scientific enterprise, considering the tendency everywhere exhibited to multiply
the resources of our country, and to find substitutes, in useless and neglected rubbish, for expensive articles employed in the arts and manufactures, that the attention of the commercial and
manufacturing public has not been directed to the
field of inquiry and research, so promising in rich
results, which the dye-lichens present.

As I have already remarked regarding the medical properties of lichens, their tinctorial qualities are equally variable. Orchil manufacturers have found, as the result of long experience, that lichens from tropical, sub-tropical, and maritime regions, are richer as a rule in colouring matter than those growing in northern latitudes and inland localities. Corticolous species, or those which grow upon trees, also yield fainter traces of colouring properties than saxicolous species, or those which grow upon rocks. Latitude, climate, temperature, moisture, exposure, elevation, nature of rocks or trees on which they grow, and in general all those conditions which affect the botanical character of the lichens, also affect their chemical character, and so render them more or less serviceable for the purposes of the colour manufacturer or dyer. The genus Roccella itself, which is the most highly esteemed and most commonly used of all the dye-lichens, is one of the most variable as regards its botanical character in the whole lichen-tribe ; there being a great many so-called species which might very well be referred to one specific type, and regarded simply as varieties. And this changeable appearance of the plant is accompanied with corresponding modifications of chemical properties. These properties also vary at different stages of growth of the same plant; the young thallus presenting, when sub-

jected to the hypochlorite of lime test, the most
beautiful reaction, while the older thallus is very
little coloured. Nylander, one of the most eminent of our recent lichenists, proposes the use of
the hypochlorite of lime test, as one of the easiest
and most satisfactory modes of separating and
distinguishing species of lichens which have been
confounded by systematists. The reaction is produced in many instances immediately upon the
thallus being touched by the reactive ; and deter-
minations perfectly exact can be made on specimens which are in a young and sterile state, and
in other respects very incomplete. The least fragment is sufficient for the verification of the beautiful chemical character which distinguishes species
in which other differences are scarcely visible. And
there is this advantage to be gained by such a test,
that it leads the student, by the differences which
are manifested chemically, to search with more attention for organic characters, which as a rule will be
found to accompany them. As a striking example
of the invaluable aid afforded by chemical reactives
in the study of lichens, the case of the common yellow wall-lichen, the *Parmelia parietina*, may be
mentioned. This species is often confounded with
young or sterile states of *Lecanora candelarea*, to
which it has a remarkable resemblance. But a
solution of hydrate of potash,—which is of equal

practical utility as a lichen-test with hypochlorite of lime,—applied to the very smallest fragment of the thallus or fructification of both species, shows instantly the difference between them. The *Lecanora candelarea* remains unchanged in colour by this reactive, while the *Parmelia parietina* becomes of a rich and beautiful purple. We can in this manner recognise at once the specific differences of these two lichens, even without opening the paper in which they may be wrapped, provided the paper be permeable by the solution of potash ; for in the one case the paper will remain uncoloured, while in the other it will be immediately stained with the characteristic purple coloration of the chrysophanic acid of the lichen under the potash test.

One of the most magnificent of our native lichens is the *Parmelia glomulifera*, which is found occasionally on mossy rocks and trees in mountainous districts. It grows in immense profusion on the aged beech trees, which form a splendid avenue leading from Inverary Castle to the Dhu Loch ; each huge tree being covered from head to foot with a cuirass of this lichen. Some of the individual specimens are very large, and bear fruit with the utmost prodigality. The plant may be known at once by its enormous size, its thick orbicular leathery thallus, and its beautifully scolloped segments, the angles of which are perfectly circu-

lar, as if stamped out with an embossing machine.
The surface of the thallus is covered with large
tufted excrescences of a dark greenish colour ;
while the apothecia occur at all stages of growth
from small round tubercles with a point in their
centre, to large concave shields about three-fourths
of an inch in diameter, of a tawny-red colour. When
dry, the lichen is of a glaucous grey colour ; but
in wet weather it assumes the most vivid green,
and contrasts beautifully with its red fructification.
In Norway I saw a magnificent lichen in the woods
of Lillehammer, and very generally throughout the
country, covering the mossy ground like our Dog
Lichen with wide-spreading rosettes of a yellowish green colour, called the *Nephroma arctica.*
Its fructification, which is of a rich chocolate colour, is on the under surface of the segments of the
thallus, while that of the closely allied Dog Lichen
is on the upper side, like the nails of the human
hand.   It is a peculiarly northern lichen, being
altogether unknown in this country.   I was also
greatly delighted in Norway with the immense
profusion of the *Cetraria juniperina* with its rich
yellow shrubby thallus, and olive-coloured fructification.   It covered all the trees in the greatest
abundance in every wood I explored, and fruited
lavishly ; while in this country it is exceedingly
scarce, and occurs only in fir-woods in a few places

in the north of Scotland, and has never been seen in fructification. It used to be employed in Norway and Sweden as a cure for jaundice, on account of its yellow colour. Another of our finest lichens is the *Sphærophoron coralloides*, which grows upon mossy rocks in sub-alpine regions, and looks not unlike the common coralline of the sea-shore. It forms shrub-like tufts of a pale brownish colour with grey branches and twigs, and produces fructification in the form of a pulverulent black ball at the summit of the principal stem. There is another coralline-like lichen, the *Dufourea retiformis* of New Zealand, perhaps the loveliest of all the lichens, which resembles a combination of the Flustra membranacea or common sea-mat of our sea-shores and the reindeer-moss. Its lace-like tufts look as if woven in Nature's finest loom. These two lichens repeat on dry land the idea of the corallines of the sea, and show how closely related are the algæ of the ocean of water to the algæ of the ocean of the atmosphere.

Speaking of this curious relationship between the cryptogamia of the land and the sea, it is obvious that lichens must have been the first land plants with which the earth was covered when it emerged from the primeval waters ; and very probably, as was the case with the fern-tribe, these primitive lichens may have attained a size and

luxuriance, of which their dwarfed modern representatives can give us no idea. If the Palæozoic was
the age of Acrogens, the Eozoic may have been that
of Thallophytes and Anophytes. Gigantic mosses
and lichens may have been the sole vegetation,
and may have produced the extensive deposits of
graphite which exist in the Lower Silurian, and
which has as yet afforded no remains of land
plants. To the chemist the presence of graphite,
or of a metallic sulphide in a rock, affords clear
evidence of the intervention of organic life ; and
these indirect evidences are met with even in the
oldest known stratified rocks. Nay, strange to
say, the presence of graphite, native iron, and sul-
phides in most aërolites, discloses the startling
fact that these bodies come from a region where
vegetable life has performed a part not unlike
that which it still plays upon our globe. Professor
Daubeny has suggested that the former existence
of vegetable life in the oldest rocks containing
no fossil remains, or even graphite, iron or sul-
phurets, may be ascertained by the presence in
them of phosphoric acid, which is essential to every
form of life, and which cannot be dissipated by
any amount of heat or metamorphic action, when
in a state of combination. The minutest traces
of this acid in a rock that might otherwise escape
notice may be detected indirectly by sowing barley

seed in a sample of the pulverized rock, and deter-
mining whether the growing plant yields more phosphoric acid than was present in the grain ; it being
evident that any excess must have been derived
from the rock from which it drew its nourishment.

As lichens are thus the earth's first mercy, so
they are its last gift to us. They 'cover with
strange and tender honour the scarred disgrace
of ruin, laying their quiet finger on the trembling
stones to teach them rest.' Nearer than the remains
of castle or hovel ; nearer than the garden trees
which they invest with a hoary reverence when all
service of fruit-bearing is over, the lichens come to
us. They take up their watch upon the tomb
that is forsaken by all else. More constant and
faithful even than the moss that fills up the hollow
inscription with its soft green velvet lines, the
very handwriting of Nature striving to keep in
remembrance what man has forgotten, the lichen
endures when the moss decays and fades away ;
and, in its living letters clinging to the worn stone,
conveys the significant lesson of immortality, of
life in the midst of death. Nowhere have I been
more struck with the last tender ministries of the
lichens than in a romantic churchyard beside a
ruined ivy-grown chapel in a little island in Loch
Leven, off Ballachulish. In that churchyard the
Macdonalds of Glencoe, who perished in the ter-

rible massacre, are buried ; and over their slate
grave-stones, weather-worn and battered, grey
and yellow lichens spread their rich halo of living
glory, obliterating heraldic symbol and pathetic
tale, and subduing the tragic memories of a
stormy period to the profound repose of sea and
mountain around. The contrast between the
lichens on the tomb and the story of bloodshed
which they veiled with hushed softness, could not
have been more striking. How eloquently did
these meek creatures speak of the peace and permanence of nature which succeed all the works and
passions of man ! Gently and slowly they were
bringing back with their ' rounded bosses of furred
and beaming green, their starred divisions of rubied
bloom, and their traceries of intricate silver, subdued and pensive and framed for simplest, sweetest
offices of grace,' the last work of man, as man himself was being brought back into the bosom of the
Universal Mother. The woods, the blossoms, the
gift-bearing grasses, as Ruskin beautifully says, had
done their part for a time ; but these lichens are
doing service for ever. Trees for the builder's
yard, flowers for the bride's chamber, corn for the
granary, lichens for the grave.

CHAPTER III.

IN these days of popular science, when the most abstruse subjects come to us in forms as light and easy as the whisperings of confidential friends, or the chit-chat of the family circle, no department of natural history is more extensively and successfully studied than that which relates to the algæ or sea-weeds. And this need not be wondered at, for there is no class of plants more interesting, whether we regard the beauty of their colours, the gracefulness and variety of their forms, or the romantic situations in which they occur. The invention of that elegant ornament of the parlour and drawing-room, the aquarium, now so popular, has afforded great facilities for the study of these plants, under conditions and circumstances closely analogous to those of their native haunts ; and much insight has· in conse-

quence been obtained into their functions and
habits, which would otherwise either remain in obscurity, or be revealed only by the chance fortune
of the hour.   It would be interesting to state some
of the novel facts thus elicited.   But this would be
irrelevant, as our attention in this chapter is to be
occupied not with the history of the algæ or sea-
weeds as a whole, but only with that distinct and
well-marked section of the family which inhabits
fresh water exclusively, whose economy is altogether peculiar, and whose forms are widely different from the lovely Plocamiums and Deles-
serias, which we frequently observe with admiration in our wanderings along the sea-shore.

   There is a peculiar charm about fresh-water
algæ, derived from the nature of the element in
which they live.   Aquatic plants of all kinds are
more interesting than land plants.   Water is so
bright, so pure, so transparent, so fit an emblem
of that spiritual element in which our souls should
bathe and be strengthened, from which they should
drink and be satisfied.   It is a perpetual baptism
of refreshment to the mind and senses.   It ideal-
izes every object in it and around it ; the commonest and most vulgar scenes, reflected in its
clear mirror, are pictorial and romantic.   It is ever
varying in its unity, so that the eye never wearies
of gazing upon it.   All these associations invest

the confervæ which flourish in it with a peculiar interest, independent of their own mysteries of structure and function. They mingle with the snow-white chalices and broad velvet leaves of the lilies, in the tranquil shallows of the moorland lake ; and, with the golden hues of the sunset, and the rosy blush of the heather-hills around, create a scene of enchantment in the clear pellucid depths. Their dishevelled tresses toss wildly in the foamy rapids of the water-fall, whose misty spray rises to freshen all the scenery around, and whose 'sound of many waters' fills the mind with a feeling of animated delight and bounding vivacity. They float in long, graceful wreaths and glossy traverses of silken change in the streamlet, wherever it clothes a jutting mass of rock with gemmed and sparkling folds of liquid drapery, and are burnished by the snowy water through every fibre into fitful brightness. They lie like motionless clouds in the blue depths of the tranquil linn, that just ripples for pleasure, as it murmurs to itself a sinless secret hidden for ever in its heart. They fringe the pebbly sides of the river, whose deep bulging fulness flows on unceasingly, ever diffusing freshness through the green pastures which it gladdens, and beneath the drooping willows and alders that gratefully murmur over it. They luxuriate in the cold clear springs which form a

N

feature of the most exquisite beauty in the bleak
Alpine scenery, gushing up in exposed and rocky
spots, and gurgling down the sides of the hills
through beds of the softest and most beautiful
moss ; not the verdant velvet which covers with a
short curling nap the ancient rock and the grey
old tree, but long slender plumes waving under
the water, and assuming through its mirror a tinge
of the brightest golden green. In gathering or
admiring these humble plants in such romantic
situations, a sense of the beauty of the Greek my-
thology is awakened in the heart, more vivid and
real than is experienced in other circumstances.
It seems easy to believe, in quiet far-off scenes
where a solitary coot sailing on the water is an interruption to the solitude, and where the link that
binds us to the common busy earth is broken and
dropped, that the dryads are still hiding among
the trees around, and the nymphs gazing upon
their own reflected beauty in the limpid stream.
The filaments of the confervæ, lying deeper in the
fountain than one's own image, look like the green
hair of the naiads ; and it requires but little exercise
of the imagination, to fill up the exquisite forms
with their zones of rainbow drops and robes of
filmy water-moss, and the beautiful, pure, passionless faces of the invisible bathers to whom the
flowing, luxuriant tresses belong.

By the fresh-water confervæ we are brought to the very boundaries of the inscrutable ; into those arcana of nature where life, reduced to its simplest expression, seems invested with a deeper and more thrilling mystery. They are the very lowest in the scale of vegetation, and approximate so closely to certain animals both in form and in vital functions, that the best naturalists are unable to draw the line of distinction between their simplest species and the humblest animal organisms, or, indeed, to determine whether they possess vitality or not. They confound and neutralize the old arbitrary definitions of the three kingdoms of nature. Neither the power of voluntary motion nor chemical composition can be called the characteristic by which they are separated from animals ; nor can mere appearance or ostensible mode of production be regarded as sufficient to distinguish them from minerals. All we can say regarding them, and regarding the animals with which they form connecting links, and into which some even say they are transmuted, being animals at one period of their lives and vegetables at another, is merely that the two lines or systems of life seem to start as it were from a common point at the base ; the inferior forms bearing a certain similarity to each other in structure and functions, which gradually disappears as we ascend the scale

of development, until at the summit we behold
those vast differences which distinguish an elephant from a palm-tree.

In this class of plants, minute and obscure
although they are, the infinite resources of Creative
power are perhaps more overwhelmingly revealed
to our perceptions, than in even the highest orders
of the vegetable kingdom. The most unwearied
research, continued for centuries, has not yet assigned limits to that amazing variety which is their
most remarkable feature, numbering as they do
species that baffle classification, and within which
a still more astounding variety of individual types
are to be found.

Every one is familiar with that green slimy
matter, which during the spring and summer
months creams over the surface of the stagnant
pool, the half dried-up streamlet, or the wayside
ditch ; but there are few who regard it otherwise
than as a disagreeable scum or impurity, to which in
Scotland the expressive name of *slaak* has been
applied. It is in reality, however, an aggregation
of plants, perfect in all their parts, and furnished
with peculiar organs of nutrition and reproduction. Let us place a small portion of it on a concave glass, containing a drop or two of water sufficient to float it freely, and then place it under
the microscope for examination, and what a

beautiful spectacle is unfolded to us! That which
to the naked eye appears a mere gelatinous mass
of shapeless filth, is found to be composed of a
thousand delicate and exquisitely formed threads
or filaments, which in some instances are simple
(Fig. 17), and in others branch, radiate, and inter-
lace like the most beautiful network. Each of
these threads is a transparent tube filled with endochrome, or little green cells, forming different

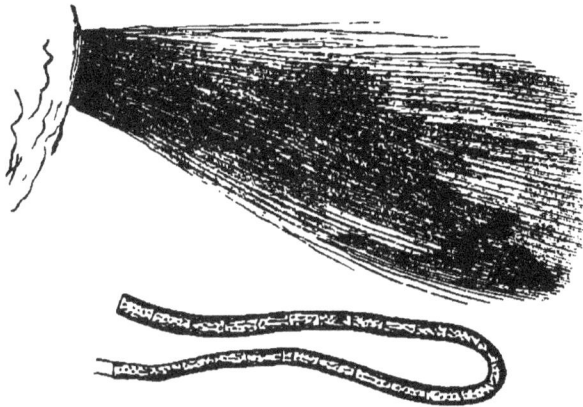

FIG. 17.—CONFERVA RIVULARIS.

figures or articulations, placed at regular intervals,
and containing minute germs floating in mucilaginous matter. This internal matter is the fructification; and may be regarded as the first and
simplest step in that long series of changes, by
which certain parts of the vegetable organism are
set apart and moulded with reference to the function of reproduction. The modes of propagation

are very diversified ; almost every family having a peculiar plan of its own. Some reproduce the species by the simple breaking up of the filaments into larger or shorter pieces, or into single joints. Others are reproduced by gonidia or zoospores developed from the contents of the filaments and covered with vibratile cilia, by means of which they swim actively in the water. But the most remarkable mode of propagation is that which is known as the process of conjugation. When two filaments approximate, each throws out from one side a small process, which unites with a corresponding process from the side of the other ; the two ends of the processes become absorbed, and the interval between the two plants is thus bridged over by a transverse tube. The endochrome of the one cell then passes through the communication thus formed into the other, and the contents of both cells become intimately mixed and form a round mass, which ultimately becomes the seeds or resting-spores by which new plants of the same kind are destined to be produced. In consequence of some differences of structure, to our eyes inappreciable, the filaments appear to exercise in one case the function of the male, and in another that of the female. But how is it, it may be asked, that process meets process in two contiguous filaments, and form between them a germinating

spore ? By what power is a plant given to understand, that a similar plant lies in its immediate
neighbourhood, ready to carry on the necessary
fructifying process ? Certainly we can consider it
nothing else than a species of the same indefinable
operation, which prompts the bee to construct a
cell of an hexagonal form, or a bird to build a nest
in the manner peculiar to its species.

We thus find that these obscure plants form no
exception to the very general, if not universal law
that each species of living being requires two distinct elements for its perpetuation. Sexual elements have been detected in most of the cryptogamic plants, and in a short time will probably be
discovered in all. The power of reproduction by
segmentation, or the production of numerous suc-
cessions of asexual fertile generations, which, in
common with many others of the humblest organisms, vegetable and animal, the confervæ possess,
is in all cases limited, the species necessarily reverting to sexual admixture for its perpetuation.
The germs produced by the conjugation of approximated individuals, when fully ripe, burst the
cells in which they are confined, and are consigned
to the surrounding water, where they float about,
until they meet with some substance to which their
mucilage enables them to adhere ; and once established in a congenial situation, they spring up into

new plants, and extend themselves with amazing
rapidity, in a week or two producing thousands
and tens of thousands of individuals. The lives
of the fresh-water algæ rarely exceed a year in
duration, many of them dying in the course of a
few months or weeks. ˙They complete the process
of reproduction early in spring, and last during
the summer, perishing in the autumn, and disap-
pearing altogether in winter. No sooner does the
ice, which had bound up the streamlet in its silent
fetters, melt under the warm rays of the sun, allowing its water to flow merrily on, and flash and
sparkle in the sunbeams, than every stone in its
bed, though brown and naked before, is suddenly,
as if by magic, invested with a green velvet coating, whose long graceful filaments float freely
with the water. Every ditch and marsh, every
rivulet, every hoof-mark and rut on the road
where water has accumulated, is filled with green
clouds of these mysterious plants. The purposes
which they serve in these situations are sufficiently
obvious. Though associated in our minds with
stagnation, putrefaction, and malaria, they are the
scavengers, the water-filters of nature. Like the
flowers and the trees, that on dry land remove the
impurities with which the animal world is continually tainting the atmosphere, they purify the
waters, by assimilating the decaying matter which

they contain ; while their own tissues form food and shelter to myriads of animalcules, that wander over these—to them—trackless fields and endless mazes, and convert the waste pools and ditches of the wayside into scenes of busy life and enjoyment. This mutual adjustment between the economies of the animal and vegetable kingdoms, whereby the vital functions of each are maintained in the utmost efficiency, is one of the most beautiful and striking phenomena of organic nature.

The largest of the fresh-water algæ is the River Lemania (*Lemania fluviatilis*), which bears a close relation to the lower Fucoids of our rocky seashores. It is never found growing in stagnant waters. Indeed, it is said to languish and die, when the streams in which it is produced have, by some cause or other, been converted into motionless pools. It loves to grow in clear swift rivers, flowing with a strong current over a rough and rocky bed, and in Alpine streamlets, on the very verge of the numerous cascades which they form during their descent from the hills. It is a matter of surprise how it can sustain the immense force and weight of the impetuous waters, without being uprooted and carried away. Examination will, however, discover that it has been wonderfully provided with means to enable it to brave the dangers to which in such situations it is exposed.

Its filaments are elastic, rigid, and bristly, from three to six inches in length, about the size of a hog's bristle, and knotted throughout at equal distances with prominent swelling joints, like those of the bamboo cane. They spring from a tough cartilaginous disk, so firmly applied to the rock as to require a very considerable force to detach it. It is impossible to convey in words, the same strong impression of fitness and perfection of contrivance, which a glance at the plant in its native haunts would produce. . It appears one of the most striking examples of that compensatory adaptation of structure to requirements, which we observe more or less in all the lowest plants ; in the moss, which, considering its size, adheres with more tenacity to its growing place than the oak of centuries, that strikes out its roots over half an acre of ground ; and in the minute crustaceous lichen, apparently as hard as the rock upon which it is produced, over which the devastating storms of the Alpine summit sweep for years without inflicting upon it the slightest injury. The colour of the Lemania, when fresh, is of a fine deep olive-green ; but it changes to black when dried and placed in the herbarium. The dilatations or gouty joints are owing to the development of the sporules within the fronds ; and these may be squeezed out by being compressed between the fingers. The force with which they naturally break through the tough and cartilaginous skin of the frond, in order to form independent individuals, is not the least curious circumstance in the economy of this strange plant. Bory, to whom we are indebted for the name, informs us that the recent filaments of the Lemania, when applied to the flame of a candle, explode and extinguish it, while a remarkable movement of retraction is felt by the fingers which hold them. The plant is rich in nitrogen, and when burnt yields ammoniacal vapours. The spores at first vegetate into slender filaments, which constitute a sort of prothallus or pro-embryo. From the cells of these filaments spring up after a time thick branchlets, which are at first wholly dependent upon the cells from which they arise ; but they soon acquire rootlets at their base, and rapidly elongating grow into the cartilaginous bristle-like tufts characteristic of the mature plant.

The confervæ generally grow in single branchless filaments, forming a loose fleecy stratum ; but sometimes they are aggregated together into singular forms. There is one species known as the water-net or water-flannel (*Hydrodictyon utriculatum*), which looks more like a piece of green baize manufactured by man, than a production of nature. It forms a beautiful tubular purse or net, with regular polygonal meshes articulated at the inter-

sections, varying from half a line to half an inch in
diameter, grey on the one side, and green on the
other. The filaments which compose these meshes
are sometimes slender as a human hair, and sometimes as coarse as a hog's bristle, feeling harsh to
the touch when handled. There is no granular
fructification within the filaments, consequently
the plant is propagated viviparously, each of the
articulations giving birth to new filaments, which
add new meshes to the net, and, in this singular
manner, a single individual often weaves a green
network covering over the whole surface of a
pond. It is not attached to any aquatic plants,
but floats freely in the water. It is rare in Scotland and Ireland, but is of common occurrence in
ponds and ditches in the middle and south of
England.

Another curious conferva, which departs widely
from the normal form, is the *Moor Ball* or *Globe
Conferva* (*Conferva ægagropila*). It is found occasionally in lakes in North Wales, in Cumber-
land, and in the Highlands of Scotland. The filaments radiating from a central point form dense
round pale-green balls, as if composed of faded
silk thread, sometimes four inches in diameter, and
having a strong resemblance to the hair balls that
are found in the stomachs of goats. They are
sometimes employed as pen-wipers in the places

where they are found. These balls float freely at a small depth in the water, and are often washed ashore by the waves, where they accumulate in dense masses, and are again covered over with a parasitic confervoid growth.

In ditches by the waysides, may often be seen large dark-green intensely slimy masses of rigid filaments as thick as horse hair. This belongs to the genus *Zygnema* (Fig. 18), one of the largest and most curious divisions of the confervæ. Under

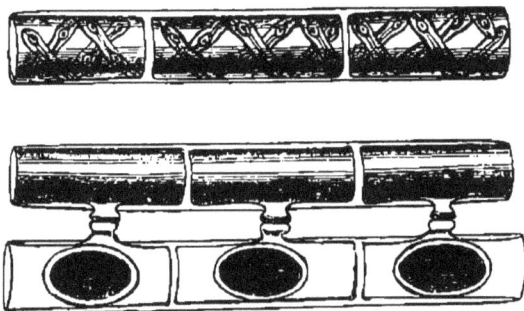

Fig. 18.—ZYGNEMA DECIMINUM.

the microscope, the filaments are found joined parallel to each other by transverse tubes, and marked by articulations longer than broad. They are remarkable for exhibiting more distinctly than in the other confervæ the process, already alluded to, of conjugation or inosculation of neighbouring filaments, in order to the production of the rest-ing-spores. They are also distinguished for the spiral arrangement of their internal granular mat-ter, which, in some cases, is like a continued multi-

plication of the Roman numeral x, and in other
cases resembles a series of the letter v ; the spiral
rings after conjugating producing a dark coloured
globule in one of the filaments.   The spiral, it may
be remarked, is the first regular form which falls under the notice of the unassisted vision, and unites
in itself the two principles of unity and variety.
In the inner surface of the cell it may be seen first
of all ; and all the parts of the plant subsequently
added, whether microscopic or visible, assume this
form.   So universal is the spiral tendency throughout the vegetable kingdom, that, beginning with
the cotyledons or seed-lobes, the whole of the
appendages of the axes of plants,—leaves, calyx,
corolla, stamens, and carpels, form in their normal
state an uninterrupted spiral, governed by laws
which are nearly constant.   It is very interesting
to trace in the obscure and humble organisms
under consideration, the order and harmony which
are so characteristic of the highest works of creation, which are in striking accordance with the
native principles of beauty implanted in the human
mind, and which proceed, we must believe, from
Infinite Wisdom.   The Zygnemas form the principal fresh-water algæ of India, occurring in pools
and streams in the central districts, as well as
among the Himalayas.   They ascend as high as
15,000 feet on these mountains, forming cloudy

masses in the ice-cold springs which trickle from
the edges of glaciers. It may be remarked that
the algæ which conjugate are found only in perfectly still water; for calmness is absolutely necessary to enable them to carry out their peculiar
method of reproduction. They adhere closely to
paper; but they lose their beautiful green tints in
drying, changing to a dull black colour.

There is a very remarkable class of confervæ
called Oscillatoriæ, on account of the singular oscil-
lating motion observed in the filaments by various naturalists, thus connecting them apparently
with the animal kingdom; the power of voluntary
motion being one of the chief characteristics distinguishing animal from vegetable life. These Oscillatoriæ grow in masses of filaments based on a
mucilaginous substance, the remains of old dead
individuals deprived of their colour and agglutin-
ated together, the whole emitting a strong odour
of sulphuretted hydrogen which is extremely disagreeable, and sometimes causes severe headache.
As a family they are distinguished for the brilliancy of their colours, their rich gradations of violet, purple, dark brown, and glossy black, and the
metallic or æruginous appearance of their shades of
green. They have been found in a great variety of
situations, ascending as high as 17,000 feet, or even
18,000 feet on the Himalayas. Some species grow

in moist, damp places, where they form a thin glossy-black pellicle of indefinite extent over the ground, strongly resembling, when dry, a piece of black satin (Fig. 19). Others are found in ditches and ponds ; a third species spreads extensively over damp walls in autumn and winter, a peculiar variety covering the damp walls in the inside of some Suffolk churches with bright sky-blue mould-like patches ; a fourth is often found on rotten timber, and trunks of aged trees where rain-water

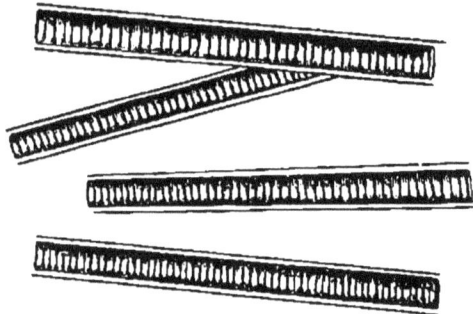

FIG. 19.—OSCILLATORIA NIGRA.

trickles down. They may be found parasitic upon mosses in rapid streams, and forming thick glossy strata of a dull-brown or vivid-green colour, at the bottom of clear, tranquil linns, wherever a film of soil is allowed to accumulate upon the naked slip-pery rocks. They are found in sulphur springs, forming pale yellow continuous tufts wherever the water retains sensible sulphureous qualities, as if the hepatic gas were necessary to their growth ; and in the celebrated warm waters of Bath, a peculiar species grows in broad velvet-like patches of a dark-green colour. Their vitality is so great that they are capable of enduring the extremes of heat and cold, for they have been found on fragments of ice in Melville Island, where the temperature is considerably below zero ; and they have been found growing like vegetable salamanders in thermal springs in different parts of the globe, where the heat is sometimes so great that the inhabitants of the surrounding districts dress their food over them, and use them for other economic purposes instead of fire. Sir Humphry Davy in his *Consolations in Travel* mentions that the floating islands of Oscillatoriæ, which are constantly found in Lake Solfatara in Italy, exhibit a striking example of the luxuriance of cryptogamic vegetation in an atmosphere impregnated with carbonic gas. They are supplied by this gas, which is constantly escaping from the bottom of the lake, with a violence that causes the water to boil.

A magnificent species forms thick woolly fleeces of a deep red colour, in the ponds and collections of water in the central and western districts of India, occurring in great profusion in the hot, sweltering valleys of the great Runjeet, ascending into Nepaul and the lower slopes of the Himalayas. The most singular member of this curious group, however, is the *Trichodesmium erythræum*

O

of Ehrenberg.. It occurs in extraordinary profusion in the Red Sea, over the surface of which it
spreads for many miles, according to the direction
of the wind, in the form of a dark-red shining
scum. It is composed of little bundles of filaments marked with striæ, which have been compared to minute fragments of chopped hay. In
certain states of the weather it emits a disagreeable, pungent smell, affecting strongly the mucous
membrane, and causing violent sneezing and oph-
thalmia, thus adding to the list of annoyances
which render the passage of the Red Sea peculiarly disagreeable to passengers from the West.
The habit of this alga is widely different from
that of its congeners, and resembles that of the
Sargassums or Gulf-weeds, which form extensive
floating meadows to the west of the Azores, and
are supposed to indicate the site of submerged
lands. The name of the Red Sea greatly puzzled
the ancients, and has occasioned in later times a
display of much superfluous learning to determine
whether it was derived from the colour of the
water, the reflection of the red coral sand-banks
and the neighbouring mountains, or the solar rays
struggling through a dense atmosphere. Another
conjecture may be hazarded, that it has acquired
its denomination from the extreme prevalence and
conspicuousness of this red alga in its waters.

The filaments of all the species of Oscillatoria are elastic, simple, exceedingly minute, and mathematically straight. They are distinguished by close parallel rings or transverse markings easily separating from each other. The motion of oscillation, for which all the species are distinguished, is in some remarkably vivid, and would favour the supposition that they are animals and not plants, were it not that their other characteristics are peculiarly those of vegetables. The filaments continually move from right to left, or from left to right, but in a very irregular manner, some going in one direction and others in another; some being at rest while others are in motion. This lateral oscillation has been attributed to various causes. The majority of naturalists, inclining to the opinion that it is mechanical and not voluntary, have ascribed it to rapidity of growth, which, in such simple plants, is excessive; to the molecular action of light, or to the agitation, by hidden causes, of the water in which the filaments are immersed for inspection. But none of these suppositions afford a satisfactory explanation, as Captain Carmichael ascertained by the following simple contrivance: He placed a small portion of the stratum of a species of Oscillatoria, composed of a great many individuals united together, in a watch-glass filled with water, and covered it with a thin plate of

mica, which effectually excluded the outer air, and
kept the water as motionless and fixed as a piece
of ice. The glass, with its contents thus arranged,
was placed under the microscope, and the oscillation of the filaments was observed most vividly,
there being no possibility of disturbance by the
agitation of the water, showing clearly that the
singular movement was independent of that cause.
'The action of light,' says this accomplished naturalist, ' as a cause of motion, cannot be directly dis-
proved, because we cannot view our specimens in
the dark ; but indirectly there is nothing easier.
If a watch-glass, charged as above, be laid aside
for a night, it will be found that by next morning
not only a considerable radiation has taken place,
but that multitudes of the filaments have entirely
escaped from the stratum, both indicating motion
independent of light. Rapidity of growth will
show itself in a prolongation of the filaments, but
will not account for this oscillation to the right
and left, and still less for their travelling in the
course of a few hours to the distance of ten times
their own length from the stratum. This last is a
kind of motion unexampled, I believe, in the vegetable kingdom.' Many species, it may be remarked, possess at their extremity a tuft of very
minute, delicate cilia, which possess the power of
imparting motion to the filaments on which they

are developed. Another strange fact in the economy of these very singular and anomalous plants is the extremely limited term of their existence. Their cycle of life is often completed in three or four days. The community of individuals associated together in one patch or stratum live for several months; but the individuals themselves die off, and are succeeded by others with a rapidity truly marvellous. The remains of the dead filaments form the bases of the living ones, and thus they go on increasing in depth and breadth until they often cover the whole bed of a streamlet. This peculiarity connects them with the coral-zoophytes, and supplies another link between the animal and vegetable kingdoms. Mr. Sorby has recently made some valuable observations in the spectrum analysis of the Oscillatorias. Ehrenberg called the lowest organizations, belonging to the fresh-water algæ, 'the milky-way of the vegetable kingdom;' and it is an interesting reflection that the same method which has yielded such wonderful results when applied to the nebulæ of telescopic research, is promising to be equally successful when applied to the nebulæ of microscopic research. By examining the spectrum of a coloured solution obtained from certain species of Oscillatoria, he has found three perfectly distinct colouring matters, which also occur in lichens,

especially the dog-lichen when growing in a damp
and shady situation, and in the olive sea-weeds.
The relative amount of these colouring matters
varies with the degree of exposure to sun and
air ; and in this respect the Oscillatorias furnish
a most interesting series of connecting links between olive sea-weeds and lichens. When their
vitality is very much reduced by want of light,
their type of colouring closely approaches that of
the olive sea-weeds ; whereas, when they are exposed to much air and light, the type approaches
to that of such lichens as the Peltidea canina or
Dog-Lichen. There is thus a striking analogy between the results due to abnormally reduced or
increased vitality in the same kind of plants, and
the normal characters of lower and higher classes
of plants.

Several obscure and curious organisms have
been included by botanists in this vast and varied
order of plants, some of which are supposed to be
fungi in an embryonic or imperfectly developed
state. They are composed of hyaline or coloured
articulated filaments, aggregated together and
forming a kind of fibrous crust, sprinkled over
with loose granules. The localities where many
of them are found prove that they are not genuine
algæ, or even independent organisms, but probably
only the mycelia or spawn of various fungi.

One curious species is found on windows and damp glass in shady places, where it forms round white spots, radiating like a spider's web from a centre, and sprinkled with minute, whitish, powdery particles. Another forms simple, transparent, club-shaped filaments, from a line to an inch in length, on the bodies of fishes and dead flies found on decaying leaves and weeds in the water. Several species are found in chemical solutions which are decomposed by them, in various infusions, such as distilled rose-water, dissolved muriate of barytes, and gum-dragon, and inorganic liquids undergoing fermentation, vinous, acetous, or putrefactive. The white flocculent matter often found on the surface of old stale ink, and the yellow hyaline filaments found at the bottom of wine bottles, are referred to this class of plants, to which the generic name of Hygrocrocis has been given, from their byssoid nature, and the situations which they affect. All these spurious substances are now excluded from the list of algæ.

There is one species of alga, the saffron rock byssus (*Chroolepis aureus*), which deserves, on account of its beauty, more than a passing notice. Unlike the other confervoid algæ, which are found in moist situations or in water, it is restricted to the shady side of overhanging cliffs, trunks of trees, leaves and other objects, and never grows in water. It is abundant in the Highlands of Scotland, in deep, leaf-embowered ravines near a mountain-lake or waterfall. It grows among dense cushions of the beautiful apple and other mosses, to which it affords a fine contrast by its velvety tufts of a deep orange colour, which are rendered especially brilliant by the stray sunbeams that chance to reach its growing place. It affects the trunks of larch and other pine trees in the Highland woods, often covering them from head to foot with its scarlet livery. I remember being greatly struck with its profusion and fine effect in the woods along the bay of Tobermory in the island of Mull. In similar localities, and particularly on the micaceous rocks on the Highland mountains, may often be observed its Ethiopian relative, the black rock byssus (*Chroolepis ebeneus*), forming a thin, black, velvety patch of indefinite extent, composed of fine, branched, black hairs, closely matted together, and sometimes sprinkled over with black powder. Few would suspect its vegetable character ; indeed, it bears a greater resemblance to a piece of black felt scraped from a hat than to any plant. Both these plants are supposed to be peculiar states of certain lichens, their reproductive bodies being very similar. A yellowish species of this genus called *C. odoratus*, popularly known as the ' sweet scented moss,' is occasionally found

upon lichens and old trees, giving them a rusty-
yellow look. A remarkable species is found
only on yew-trees of great age, growing in the
deep clefts near the root, from which the juice of
the tree exudes. Fed by that sap it increases from
a black filamentous crust, to a thick corky substance, which exhibits when broken a series of concentric layers which indicate its age. When well
dried it takes fire very readily from a spark, and
burns like tinder. This singular substance, which
is unlike any other plant, bears a greater resemblance to a fungus than to an alga. I have
gathered it on old yew-trees at Cleish Castle in
Kinross-shire, and beside a ruined castle of the
Macfarlanes in an island near the upper end of
Loch Lomond. Upon the base of the abbot Mac-
kinnon's tomb in the ruined abbey of Iona, and
abundant in Fingal's Cave at Staffa, is found
the purple *Trentepohlia*, which looks like a piece
of crimson plush or velvet, and exhales when
moistened a sweet scent somewhat like that of
violets. But the loveliest species of this family is
T. pulchella, which grows on naked rocks or on
aquatic mosses in mountain streams. It is of a
rich violet or carmine colour, and imparts its own
hue to whatever it comes in contact with. The
stones and mosses in the bed of the stream where it
occurs, look as if a dyer's vat had been emptied

there. It is frequent in streams in Ireland ; but in England and Scotland it is somewhat rare and local.

The extraordinary phenomenon of red snow has long been familiarly known to scientific men in this and other countries, and has naturally enough excited the greatest interest. This singular colour in a substance with which we are accustoned to associate ideas of spotless purity and radiant whiteness, has been ascertained to result from an immense aggregation of minute plants belonging to the family now under consideration. They form the species called *Protococcus nivalis* (Fig. 20), in

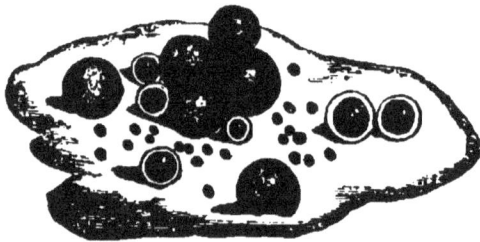

FIG. 20.—PROTOCOCCUS NIVALIS.

allusion to the extreme primitiveness of its organization, and the peculiar nature of its habitat. If we place a portion of the snow coloured with this plant upon a piece of white paper, and allow it to melt and evaporate, we find a residuum of granules just sufficient to give a faint crimson tinge to the paper. Placed under the microscope these granules resolve themselves into spherical purple cells, from the $\frac{1}{1000}$th to the $\frac{1}{3000}$th part of an inch in diameter. Each of these cells has an opening surrounded by serrated or indented lines, whose smallest diameter measures only the $\frac{1}{5000}$th part of an inch. It encloses nitrogenous contents, is tinged with chlorophyll, and contains starch. The plant, when perfect, bears no inapt resemblance to a red-currant berry ; as it decays, the red colouring matter gradually fades into a deep orange, which finally appears to change into a brown hue. The thickness of the wall of the cell does not exceed the $\frac{1}{20000}$th part of an inch. Each one of the cells may be regarded as a distinct individual plant, since it is perfectly independent of others with which it may be ·aggregated, and performs for and by itself all the functions of growth and reproduction ; having a containing membrane which absorbs liquids and gases from the surrounding matrix or elements, a contained fluid of peculiar character formed out of these materials, and a number of excessively minute granules equivalent to spores, or, as some would say, to cellular buds, which are to become the germs of new plants. There is something extremely mysterious in the performance of these widely different functions, by an organism which appears so excessively simple. That one and the same primitive cell should thus minister equally to absorption, nutri-

tion, and reproduction, is an extraordinary illustration of the fact, that the smallest and simplest
organized object is in itself, and, for the part it
was created to perform in the operations of nature,
as admirably adapted as the largest and most
complicated.

Saussure, the celebrated geologist, appears to
have been the first scientific person who noticed
this production, for in his *Voyages dans les Alpes*
he states that he found considerable patches of it
on the summit of the Brevent at Chamounix after
a fall of snow, so long ago as the year 1760, and
afterwards very frequently and in great abundance
in his wanderings over the Pennine Alps, and particularly on the Col du Géant. After this period
several eminent botanists collected it in various
places ; Ramond on the snow-capped peaks of
the Pyrenees, and Sommerfeldt on the Dovrefjeld
and other lofty hills in Norway. In March 1808,
red or rather rose-coloured snow fell in considerable quantities in the Tyrol, and on the mountains of Carinthia in Illyria; and over Carnia,
Cadore, Belluno, and Feltre, to such an extent
that the hills were covered with it to the depth of
six feet. Ten years later, it is recorded that
enormous quantities of the same substance were
spread like a bloody pall over the Apennines and
the other Italian hills, occasioning no small alarm

among the superstitious inhabitants of the surrounding districts, who looked upon it as an omen
of impending calamity, and sought refuge from
their fears in various protective ceremonies.
Among the Peruvian mountains, Darwin relates
that on several patches of snow he found this
curious appearance. His attention was called to
it by observing the footsteps of the mules stained
a pale red, as if their hoofs had been slightly
bloody. The snow was coloured only where it
had thawed very rapidly, or had been accidentally crushed.[1]   It is in the Arctic regions, however, that the red snow is found most frequently,
and in the greatest luxuriance. Sir John Ross,
during his memorable expedition to these regions
in 1808, found on the 16th of June, in about latitude 75°, a range of cliffs rising about 800 feet
above the level of the sea, and extending eight
miles in length, entirely covered with snow, which
seemed as though it had been watered by some
crimson decoction. Sir W. E. Parry found the
same phenomenon, during his heroic attempt to
reach the Pole by travelling over the ice in 1827 ;
and ascertained besides, that wherever the surface

---

[1] It is a curious circumstance, that Dr. Hooker never met with
a single specimen of red snow, during all his wanderings over the
lofty snow surfaces of the Sikkim Himalayas, especially as on
almost every mountain range elevated above the line of perpetual
snow, it has been seen, often in abundance.

of the snow-plain, although previously of its ordinary spotless hue, was crushed by the pressure
of the sledges and of the footsteps of the party,
blood-like stains appeared most visibly ; the impressions being sometimes tinged with an orange
colour, and sometimes appearing of a pale salmon
hue.

Red snow, however, seems by no means peculiar
to the Arctic regions, or the highly elevated mountains of the globe.   It has been discovered spreading over decayed leaves and mosses on the borders of small lakes, and in water tanks in hot-
houses ; and in greater perfection on limestone
rocks within reach of the spray of the ocean in
Lismore, an island off the coast of Argyllshire.
Professor Harvey, the distinguished Irish botanist,
found small patches on micaceous schist near Mill-
town Malbay, on calcareous rocks at Limerick,
and in the neighbourhood of Dublin on granite,
with only an occasional supply of moisture.   On
Ben Nevis and Ben Lawers I have more than once
detected specimens, upon the surface of the large
masses of unmelted snow, with which the summits
of these mountains are sometimes covered even in
the depth of summer.

The fact that the red snow is capable of growing in such spots as those in which it has chiefly
been found in Britain, namely, on rocks, leaves,

and mosses, exposed to occasional or frequent
inundations of water, seems to prove that the ice-
plains of the Arctic regions, and the snow-crowned
sides and summits of the European mountains,
are not its natural situations. When, however, its
germs have once been deposited in these barren
and cheerless localities, the simplicity of its organization, and the consequent strong persistency
of the vital principle in it, enable it effectually to
resist the cold ; and with that extraordinary power
of rapid development which characterizes in a
greater or less degree all the members of the
family to which it belongs, it forms in a few years,
when nourished by the moisture produced by the
melting of the icy snow during summer, vast and
dense masses, sometimes twelve feet in depth, and
extending many miles in length, which afford by
their strange contrast to the painful uniformity of
the pure and dazzling whiteness all around, a
sight more surprising to the Arctic or the Alpine
traveller than would be the realization of all the
fabled wonders of the Arabian tales.

Another supposed species of Protococcus was
discovered by Baron Wrangel in the province of
Nerike or Nericia in Sweden, not far from the
town of Orebo, and named by him *Lepraria Ker-
mesina.* The same plant was afterwards found by
various continental botanists among the fissures

of rocks, and on the under surfaces of stones in various localities, and called by them *Protococcus viridis*, or green snow. It was also observed by Martins in similar situations in Spitzbergen. It is now, however, ascertained beyond doubt to be a mere variety of *Protococcus nivalis*, as it is identical with it in every respect save colour; and this difference is owing .to the different circumstances in which it is developed. The actinic power of the solar light, aided by some peculiar, and as yet unknown property belonging to the natural whiteness of the snow itself, is highly essential in the production of the beautiful crimson or rose-colour, by which the red snow is distinguished ; but this colour, as in the case of the varieties mentioned above, gradually changes to green when secluded from the direct action of light, and developed on dark or opaque objects.

Another extremely curious plant closely allied to the red snow is the *Palmella cruenta* (Fig. 21) or Gory Dew. Like the Protococcus it consists of a number of aggregated globose cells, forming a very thin crust-like frond of a dark blood colour. Each of the cells divides first into two, then into four parts, each capable of propagating the plant. It grows on damp limestone in the open air, or on whitewashed walls, particularly in cellars, and the mouldering rooms of old neglected buildings, and figures largely in the history of the superstitions of the middle ages. Pitarello, a peasant residing at Laguaro, near Padua, observed large patches of it covering the walls of an old and rarely visited room in his house, which so closely resembled huge clots of venous blood, that the greatest curiosity and consternation were excited. The streets of Padua leading to Laguaro were thronged by anxious crowds hastening to inspect the phenomenon, and

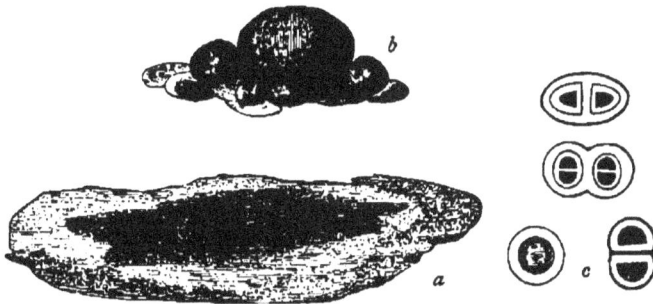

FIG. 21.—PALMELLA CRUENTA.

(*a*) Fructification slightly magnified.
(*b*) Fructification much magnified.
(*c*) Fructification highly magnified. Cells dividing into two, and then into
    four parts, each capable of propagating the plant.

full of the calamities it foreboded. Many regarded it as a direct judgment of God upon the unhappy peasant, for having forestalled corn during the famine. During the last invasion of epidemic cholera, the same plant was found in abundance, purpling the ground near Oxford, as if red wine or blood had been poured out.

In connexion with the present subject, it may

P

be interesting to glance over the several examples
of blood-prodigies which history furnishes. The
almost unanimous judgment of modern times has
stamped these examples as pure fables ; but I
think it is easy to account for the presence in
them of so much that seems incredible, and to
show how that into which the apparently fabulous
enters in so large a proportion, can yet be received
in the main as true history. Our present investigations will go far to evince that the great bulk of
what ancient writers hand down to us as prodigies
and miracles, is capable of explanation on grounds
intelligible to any ordinary understanding ; and
thus that history, so far as these things are concerned, may be true in its narrative of facts, though
it be often in error in the view it takes of the
nature of the facts narrated. That rivers have run
blood, that skies have rained blood, that the very
bread in men's houses have been sprinkled with
blood, and thus ministered death instead of nourishment to those who have eaten it, and that consecrated wafers and priestly vestments have repeatedly exhibited these horrible appearances,—
that all these wonderful things have really happened, we have every reason to believe, from the
circumstantial accounts of them given in records
purporting to be authentic, received as such by
the age that produced them, and preserved and

handed down as such to our own times. We believe the facts ; but we do not believe the explanation given of them, or the inferences deduced from
them ; our superior scientific knowledge enabling
us to account, on natural grounds, for what, in an
age of ignorance and superstition, appeared prodigies of fatal presage.

Instances of such phenomena are to be met with
in the earliest literary remains. Homer tells us
in the *Iliad* that before the death of Sarpedon

> ' The weeping heavens distilled
> A shower of blood o'er all the fatal field.'

During the first great plague of Rome, in the reign
of Romulus, we read in Plutarch that it seemed to
rain blood ; and also that when Flaminius and
Furius were leading an army against the Isubrians,
the river which ran through the Picene was seen
flowing with blood. Livy relates that the Alban
water flowed in a bloody stream, and that this and
other prodigies were expiated by sacrifices. Quintus
Curtius mentions the occurrence of blood-rain near
the end of the siege of Tyre by Alexander the
Great, staining the bread of the Macedonian
soldiers as well as of the besieged Tyrians, but
frightening the latter into surrender. Ehrenberg
has given a very full and exhaustive account of
these curious portents, from which I shall give a few
selections. In the middle of the ninth century, red·

dust and matter like coagulated blood fell from the
heavens at Constantinople.   In 1416 red rain fell
in Bohemia.   In 1501, according to several chroni-
clers, a shower of blood fell at different places.
Appearances of blood flowing from bread when
bitten, are recorded as occurring at Tours in 583 ;
at Spires in 1104 ; at Namur in 1193 ; at Rochelle
in 1163, and at many other places.   At Augsburg,
in 1199, a person having kept the consecrated
wafer in his mouth, brought it at a later period to
the priest changed into flesh and blood.   Pilgrim-
ages were not unfrequently made to witness bleeding hosts, as that of Doberan in 1201 ; and that of
Balitz, near Berlin, which had been sacrilegiously
sold by a girl to a Jew.   In 1296, the Jews at
Rotil, near Frankfort, having been reported to
have caused a host to bleed which they had
bought, a fanatical persecution of these people
took place, whereby 10,000 were said to have been
slaughtered.   Several Jews were burned at Güs-
trow, in Mecklenburg, for a similar offence.   In
1492, a priest, called Peter Dove, residing in Mecklenburg, sold two hosts to a Jew for the purpose of
redeeming a pawn ; and they having pierced them,
abundance of blood flowed out.   The priest, now
tormented with remorse, confessed the transaction,
and betrayed the Jews ; twenty of their number
were burned on an eminence at Sternberg, since

called Judenberg. In 1510, thirty-eight Jews were executed and then burned, for 'having tormented a consecrated host until the blood came.' The bleeding of the host, produced in consequence of the scepticism of the officiating priest, gave rise to the miracle of Bolsena in 1264 ; the priest's garment stained with blood being preserved until quite recent times as a relic. This gave rise to the foundation of the festival of the Corpus Christi by Urban IV., although Raphael, painting his celebrated picture in 1512, substitutes Julius II. On November 6, 1548, a red substance like coagulated blood fell with a meteor at Thuringia. In 1560, on the day of Pentecost, red rain fell at Louvaine, and on the 24th of December of the same year at Lillebonne. In 1591 a shower of blood fell at Orleans ; in 1618 in Styria ; in 1638 at Tourney ; on October 6, 1640, at Brussels ; during May 5 and 6, 1711, at Orsion in Sweden ; in 1744, near Geneva ; and in 1763 at Cleves, Utrecht, and several other places. Cloths covered with blood called *Lepra vestium* accompanied the plagues of the sixth and tenth centuries, and in the epidemics of 1500 and 1503 occasioned great alarm owing to the sign of the cross being recognised in the spots.

Dr. D'Aubigné, in his *History of the Reformation*, thus describes from the writings of Zwingle the appearance of a similar phenomenon :—'On the 26th of July, a widow, chancing to be alone in her house, in the village of Castelenschloss, suddenly beholds a frightful spectacle, blood springing from the earth all around her ; she rushes in alarm into the cottage . . . but oh, horrible! blood is flowing everywhere, from the earth, from the wainscot, and from the stones ; it falls in a stream from a basin on a shelf, and even the child's cradle overflows with it. The woman imagines that the invisible hand of an assassin has been at work, and rushes in distraction out of doors, crying, "Murder, murder!" The villagers and the monks of a neighbouring convent assemble at the noise ; they partly succeed in effacing the bloody stains ; but a little later in the day the other inhabitants of the house, sitting down in terror to eat their evening meal under the projecting eaves, suddenly discover blood bubbling up in a pond, blood flowing from the loft, blood covering all the walls of the house. Blood, blood! everywhere blood! The bailiff of Schenkenberg and the pastor of Dalheim arrive, inquire into the matter, and immediately report it to the Lords of Berne and to Zwingle.'

This extraordinary and alarming effusion of blood, along with the previously mentioned instances of bleeding hosts, and showers of blood, although plainly exaggerated by the dilated eye of fear, which in those troubled times saw everywhere frightful portents, apparently foreboding the most calamitous events, were no doubt owing to the excessive development, under peculiarly favourable circumstances, of an exceedingly minute alga, bearing a strong superficial resemblance to the red snow plant. This alga was called by Ehrenberg the purple monad, under the impression that it was an animalcule. More accurate researches, however, have since determined its vegetable nature, and it is now called *Palmella prodigiosa*, from the wonderful rapidity with which it develops and extends itself. The body of this curious atom is but from the one three-thousandth, to the one eight-thousandth of a line (twelfth of an inch) in length. Like the red snow plant, it first of all appears in the form of small, bright, red points, like so many coloured minute dewdrops, or the roe of fishes, composed of inconceivable myriads of individuals, which afterwards unite into large red-currant-jelly-like patches, coalescing and penetrating the substances upon which they are produced. Its peculiar habit would seem rather to indicate affinity with the fungi than with the algæ. Indeed by some authors it is now considered a fungus, and classed among these plants.

The accounts of blood-prodigies found in ancient history, are matched by well-authenticated phenomena which have presented themselves

within the memory of many now living. So late
as the beginning of this century, the excessive
growth of red algæ on the surface of the Elbe
made that river for several days seem to run
blood ; while shortly afterwards some portions of
the Nile reddened in the same way, and remained
blood-like and putrid for many months, thus re-
peating as it were the old miraculous plague of
Egypt. In Silliman's *North American Journal*
there appeared several years ago a description of an
extraordinary fountain of blood discovered in South
America. A person approaching the grotto from
which the waters flowed observed a disagreeable
odour, and when it was reached, he saw several pools
of the blood in a state of coagulation. Dogs ate
it eagerly. The late Don Raphael Osijo undertook
to send some bottles of this singular liquid—rivalling the famous blood of St. Januarius—to London
for analysis, but it corrupted within twenty-four
hours, bursting the bottles. Before the potato-
blight broke out in 1846, red mould spots appeared on wet linen surfaces exposed to the air in
bleaching-greens, as well as on household linens
kept in damp places in Ireland. In September
1848, Dr. Eckard, of Berlin, while attending a
cholera patient, observed the same appearance on
a plate of potatoes which had been placed in a
cupboard of the patient's house. The potatoes

were transmitted for examination to Ehrenberg, who found the colouring matter to consist of extremely minute algæ, or animalcules as he called them, somewhat allied to the *Palmella prodigiosa.* A similar appearance was observed on bread in Philadelphia in 1832 when the cholera was prevailing in that city. In the spring of the year 1825, the waters of the Lake of Morat presented an appearance in many places of being coloured with blood, and popular attention was speedily directed to this strange occurrence. M. de Candolle, however, proved that the phenomenon in question was caused by the development of myriads of the purple conferva (*Oscillatoria rubescens*). The phenomenon occurred every spring for several years, when the fishermen of the neighbourhood, more poetical than this class of persons usually are, remarked that 'the lake was in flower.' M. Montagne records a similar phenomenon in the *Comptes Rendus.* He happened to be at the Château du Parquet in July 1852, when the temperature had been exceedingly high for about ten successive days. This continued warmth of the atmosphere was probably instrumental in providing the conditions suitable for the development of a red parasite, which attacked all kinds of alimentary substances, and particularly pastry, imparting to them a bright

red colour, resembling arterial blood. 'The
servants,' he observes, 'much astonished at what
they saw, brought us half a fowl roasted the
previous evening, which was literally covered with
a gelatinous layer of a very intense carmine-red,
and only of a bright rose colour where the layer
was thinner. A cut melon also presented some
traces of it. Some cooked cauliflower which had
been thrown away, and which I did not see, also,
according to the people of the house, presented
the same appearance. Lastly, three days afterwards, the leg of a fowl was also attacked by the
same production.' From a microscopic examination, M. Montagne concluded it to be the same
thing as described by Ehrenberg. The particles
of which it is composed have an active molecular
motion, and hence Ehrenberg's mistake in suppos-
ing it to be an animalcule. Its resemblance to
the gelatinous specks which occur on mouldy
paste, or raw meat in an incipient state of decomposition, would seem to indicate that it is a
fungus allied to the moulds, and not an alga. Its
vitality is not impaired by desiccation, even at a
high temperature. A portion of paste containing
this Palmella was dried in an oven for forty-eight
hours, until nearly baked into biscuit, yet fragments of it readily grew when scattered on fresh-
made dough.

A red colour, closely resembling blood, not unfrequently astonishes the sailor in some parts of the ocean. Captain Tuckey mentions that the water of the Gulf of California is reddish, whence it is sometimes called the Vermilion Sea. Captain Colnett, in his interesting voyages, states that 'the set of the currents on the coast of Chili may at all times be known by noticing the direction of the beds of small blubber (gelatinous algæ) with which the coast abounds, and from which the water derives a colour like that of blood. I have often been engaged,' he adds, 'for a whole day in passing through various sets of them.' D'Orbigny also remarks that there are immense tracts off the coast of Brazil filled with small animals so numerous as to impart a red colour to the sea ; large portions are thus highly coloured, and receive from the sailors the name of the Brazil bank, which extends over a great part of the coast of the country, keeping at nearly the same distance from the shore. Another bank of the same kind occurs near Cape Horn in latitude 57°, and was encountered by Captain Cook during his third voyage. Mr. Scoresby narrates that he noticed in his last expedition to the Arctic regions in 1823, some insulated patches of reddish brown water, which were found to be occasioned by minute algæ ; and often too were the floating icebergs tinged with them of a carmine or deep orange hue. Ehrenberg frequently observed in the steppes of Siberia, lakes and other collections of water filled with red algæ. 'In a fen,' he remarks, 'with a pool of water, the dark-red blood colour was very striking even at a distance. This colour I found on examination was confined to the slimy surface, which in different places formed a shining skin. The red colour was darkest at the edge of the marsh.' How many a wonderful fairy tale has science divested of its gilded ornaments, and converted into hard fact and unvarnished truth! And how many a phenomenon, magnified by the unthinking ignorance and credulity of vulgar superstition into an evidence of supernatural agency, and an omen of future calamity, has the microscope resolved into a mere collection of minute and simple vegetables, or equally harmless animalcules!

There is a startling thought suggested by these accounts of blood-prodigies. Occurring as most of them did before the outbreak of epidemics which they were supposed to herald, and on account of which they were called *signacula*, they obviously point to the conclusion that they were developed by abnormal conditions of the atmosphere. In ordinary circumstances, but few either of the animals or plants which caused these alarming appearances are produced, and then only in obscure and isolated localities; but their seeds lie around us in immense profusion, waiting but the recurrence of similar atmospheric conditions as existed in former times, to exhibit as extraordinary a development. For all we know there may be existing amongst us the germs of other forms of life, ready to develop themselves into new manifestations of the power and wisdom of God, if it should please Him to adapt the vital envelope of our globe to the uses of other occupants. The present chemical condition of the air is admirably adapted for the healthy development of the forms of life that now exist in it; and so likewise is the water for the organisms that pervade it. But who can tell what species of plants and animals would succeed the present species, were there but the smallest change effected in the proportions of the constituents of these elements? Geology reveals to us the singular fact that, when the air and the water were densely impregnated with carbonic gas during the coal era, an extraordinary development of the humblest forms of animal and plant life was the result. The earth was covered with dense forests of ferns and mosses, and the waters were peopled with myriads of corallines. And were similar conditions of the atmosphere and the water to occur again, or should any change be produced in the existing conditions, the change, while it would prove fatal to the most highly organized of the present race of animals and plants, would stimulate into excessive growth and profusion animals and plants of the simplest construction, which are now kept in check, and occupy but the most obscure and subordinate positions in the ranks of nature's agencies. And if the advent of wide-spread plagues in the middle ages was heralded by the vast development of the confervæ and infusoria, we are led by a cogent induction to conclude that it is a change of the air and water which breeds the epidemic, and that these are the first growths of that new animal and vegetable kingdom which would succeed the existing forms, if mankind were to be swept away.

The subdivision of the confervæ to which the red snow and the gory-dew belong, contains the simplest of all vegetable forms, if, indeed, they be plants at all, occurring in shapeless gelatinous masses of all hues, covering irrigated perpendicular cliffs in dark and shady places, or rocks exposed to the spray of waterfalls, and frequently hanging down in flakes from their surface. Their extreme simplicity is more puzzling to the botanist than any amount of complexity would

have been. Their fundamental structure, in
almost all cases, appears to be simply a mass
of cells variously arranged in a jelly-like poly-
morphous substance, to which the name of frond
has been applied, more for the sake of convenience than from any sense of its propriety;
each cell being a distinct individual plant, apparently having no connexion with the other cells
to which it is placed in juxtaposition, and performing for and by itself all the processes of
nutrition and reproduction. The question naturally arises, whether these obscure and extremely
simple organisms which stand at the very lowest
extremity of the vegetable kingdom, be really
perfect plants, or rather the commencement, the
first of the transitional stages of more highly
organized plants, unable to develop themselves
owing to their being placed in unfavourable
circumstances? Some eminent botanists have
contended that the spore germs of the lower
cryptogamic plants are in all cases precisely the
same, developing themselves into different plants
according to the medium and the circumstances
in which they are placed: becoming palmellas
when produced on moist rocks, confervæ in
streams, confervoid mosses on shady banks and
fields, lichens on dry rocks when stimulated by
the action of light, and fungi when produced on

decaying substances, and excluded from air and light. And this opinion seems to be strengthened by the fact so well known to botanists, that the permanent organization of the lowest plants is very frequently only the temporary or transitional condition of higher, and that so close is the resemblance between them that without due care in watching the progress of their development, they may easily be set down as distinct species. To this theory of development, however, plausible though it looks, I do not subscribe. Some of these productions may not be autonomous, some may seem to pass into each other by intermediate forms, and may bear a close resemblance to the primordial stages of plants belonging to other tribes ; but still there are real species among these lower genera—species which are permanent and do not undergo any further transformation, for in the circumstances in which they are found they can exist and multiply and perfect their fructification independently. Few objects are more beautiful and interesting under the microscope than some of these obscure bodies, and their study is absolutely necessary to the physiologist, if he wishes to obtain a clear insight into the real character and phenomena of growth and reproduction in the higher tribes of plants, and especially the changes which take place during the very early or embryonic condition of the more complicated structures.

The *Nostoc* (Fig. 22), one of the species belonging to this strange class of plants, is interesting on account of the historical associations connected with it. It occurs in the form of a greenish jelly or slimy mass on gravelly soils, rocks, pastures, and roadsides, among grass and moss, especially in moist weather. It is widely distributed, occurring as far south as the Antarctic regions, several species having been found by Dr. Hooker on wet

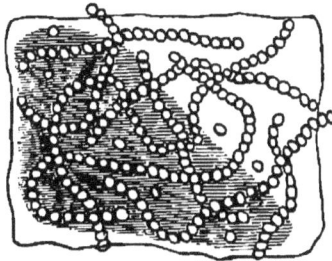

FIG. 22.—NOSTOC COMMUNE.

rocks near the sea in Kerguelen's Land. It ranges, on the other hand, as far north as Baffin's Bay, and the shores of the Polar Ocean, growing on the soft, boggy slopes of the sea-shore, from whence it is drifted about by the wind in detached masses, and forming the only vegetable production of any importance over many square leagues. Dr. Sutherland, in his fascinating journal, relates that it has often been found in great abundance on floating icebergs, and in small depressions in

Q

the snow upon the ice, at a distance of many miles from the land. It affords a welcome food,—far more palatable than the Tripe de roche, the only other edible substance which occurs in these inhospitable regions,—consisting as it does of a modification of cellulose, without any deleterious mixture. It affords food and shelter to several species of Poduræ, and an interesting little spider called *Desoria Arctica*. In the warm springs of India the Nostoc frequently occurs, and is successfully employed by the natives as an outward application for scrofulous affections, owing to the presence in it of minute quantities of an alkaline iodide. In China it is a frequent denizen of ponds and streams, whence it is carefully gathered and dried, to form an ingredient in the famous soup made of edible bird's nests. In the salt lakes of Thibet, and the marshes in the woods of New Zealand, it attains frequently gigantic proportions, forming masses of quaking gelatine, many feet in circumference. Country people suppose the Nostoc to be the remains of a fallen star, or of a Will-of-the-wisp, and hence they call it *sky* or *star-jelly*, and attach many superstitious ideas to it. Cowley in his poem on *Reason* graphically alludes to this strange fancy. It derived its name from Paracelsus, who employed it owing to its simple structure in the composition of the elixir vitæ. We find frequent mention of it in the writings of the alchemists, by whom it was highly esteemed on account of the mysterious virtues which it was supposed to possess.

The structure of this plant, simple as it appears, is very curious and interesting. Examined under the microscope, it is found to consist of a number of slender moniliform threads or necklaces of spores, invested with a firm and copious gelatine, which originated at an early stage from each individual thread, but has now become the common envelope of the whole mass. The plant is propagated by the division of these threads into their individual joints, which burst through the common jelly, and become dispersed in the water, where they are endowed with spontaneous motion, enabling them to contend against currents. These fragmentary threads divide longitudinally, at last constituting a bundle of new threads, which gradually, by increase of the gelatinous elements, assume the normal form of the species. Larger globular cells called sporanges, each producing one resting-spore which breaks out from it in germination, occur at intervals in the filaments, with others devoid of endochrome.

Another allied species is the mountain dulse of the Scotch (*Palmella montana*), occurring very frequently in patches of a deep but dull purple colour, in moist, stony places, on the mountains of

Skye, Arran, and on the west coast of Scotland, where it is used by the Highlanders, when rubbed between their hands in water, as a paste with which to purge their calves. Attached to aquatic plants, and stones at the bottom of ponds, and in the shallow margins of still lakes, may often be seen a very curious little plant belonging to this tribe called *Rivularia angulosa.* It closely resembles green-gage plums in size, shape, and appearance, and is always found associated in little colonies. It is a simple, roundish mass of gelatine, filled internally with beautiful beaded filaments. The least touch detaches it from its growing-place, when it rises to the surface of the water with the velocity of an air-bubble, and refuses to sink again, floating freely about. The whole genus Rivularia is composed of exceedingly curious plants, most of them occurring in shallow rivulets, and alpine cascades and streamlets, where they adhere, in the form of glossy bead-like dots of a dark-olive colour, to the stones of the bottom, generally preferring pure white quartz and glittering mica schist. The whole plant is not larger than a pin's-head, or a small pea; but it sometimes spreads widely in favourable situations, covering all the stones in the bed of a streamlet, and giving them an appearance, as the little, bustling, transparent waves roll and sparkle over them, as if they were full of eyes.

But the most beautiful and interesting of all the members of the gelatinous confervæ is the *Batrachospermum moniliforme* (Fig. 23), which is universally distributed over Great Britain, and is especially abundant in subalpine streamlets. It is easily known by its growing in clusters composed of branching filaments, which appear even to the naked eye like necklaces or strings of small beads, being strung, as it were, with numerous

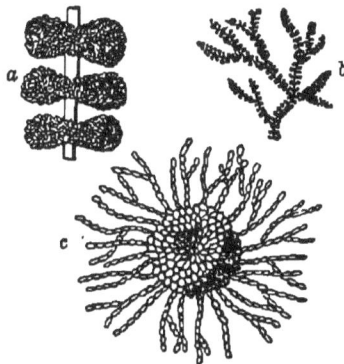

FIG. 23.—BATRACHOSPERMUM MONILIFORME.
(*a*) Magnified. (*b*) Natural size. (*c*) Section highly magnified.

gelatinous globules placed close beside each other. These branches are so exceedingly flexible that they obey the slightest movement of the water, and it is impossible to express the pleasure which is excited in the mind of the botanist, while contemplating a cluster of this little alga, in those pure, clear, sunny wells, with which he sometimes meets in his wanderings among the hills, springing up far away in lonely spots, where the curlew

builds her nest among the rushes, their mossy
sides starred with the large, snow-white flowers
of the Grass of Parnassus, and adorned with the
closed hoods and diamond-studded leaves of the
sun-dew. Every movement of the tiny fairy exem-
plifies the curve of beauty ; every filament winds
ceaselessly and rapidly through a thousand forms
of matchless grace. When removed from the·
water, however, the filaments lose all trace of organization, and slip through the fingers like a
piece of jelly or frog-spawn. The Batrachospermum occurs in the Ganges, in North America,
Hermite Island near Cape Horn, and New Zealand, and is probably distributed all over the
world. Its colour is very varied, being purple,
violet, green, yellow, and dusky brown. One
curious circumstance in this plant is, that from
the basal cells of the branches, secondary branches
grow down the stem forming a kind of rind over
it, and thus making that compound which was
originally simple. The Batrachospermum is exceedingly tenacious of life. It may be removed
with the stone to which it adheres to the coldest
or the warmest water, and it will continue to live.
It may be immersed for a short time even in boiling water without destroying its vitality. Even
when dried in the herbarium for a considerable
time, if placed in water it will vegetate as before.

A remarkable plant allied to it, called *Thorea ramosissima*, occurs in the Seine and in several other rivers on the continent, and has been found in this country in the Thames. It is attached during June and July to the bottom of boats, to stones and weeds. The whole plant, which is often many feet in length, is covered with a fine mucous down. Its colour is dark green when growing in the water; but it assumes a scarlet-tint when pressed on paper, staining pieces of cotton with it, and it settles down into a fine dark purple colour when perfectly dry.

On shady walls and thatched roofs, at the foot of rocks and houses in damp situations, may often be seen a stratum of densely crowded transparent green leaves, plaited and wrinkled with rounded lobes. This plant, called *Ulva crispa*, is the terrestrial variety of the familiar green laver of the sea-coast. Another species of the same genus (*U. bulbosa*), occasionally fills stagnant pools and ditches of fresh water, with its excessively soft and lubricous masses, appearing as if in a state of fermentation. It is exactly the counterpart of the common sea species. The *Enteromorpha intestinalis*, with which every visitor to the sea-coast is familiar, adding greatly to the beauty of rocky pools, left full by the receding tide, also occurs not unfrequently in fresh-water ponds and stagnant waters in spring and summer. An allied species, *Tetraspora lubrica*, forming irregular masses of considerable extent, and exceedingly lubricous, in gently running water, has its fruit, consisting of minute granules imbedded in the fronds, loosely arranged in fours. The first stages of all these fresh-water representatives of the marine Ulvæ, are in all respects simple confervæ ; but the cells at the extremity of the filaments divide, and by repeated division, these filaments are laterally expanded, until they form a plain leaf-like frond, as in Ulva, or close all round after they have expanded, until they produce a tube or sac as in Enteromorpha.

One of the most singular of the confervaceous algæ is the *Botrydium granulatum.* It grows on the ground in moist shady situations in spring and autumn, and is perhaps more frequent than is supposed, its minute size causing it to be over-looked. It consists of a number of green vesicles of the size of a pin's head, aggregated together, and sunk, as it were, into the soil, the whole bearing a close resemblance to a miniature branch of unripe grapes, whence the name. Under the microscope, each vesicle appears filled with a watery fluid containing minute granules, which escape when ripe by an opening at the top ; in dry weather the upper part collapses, sinks in, and becomes cup-shaped. The vesicles are attached to the soil by a tuft of root-like fibres, into which their fluid contents descend when pressed. This singular provision is necessary, as the plant is frequently exposed to dry air, which absorbs the moisture on the surface of its native soil, and would consequently wither it, were it not furnished with radicles, which penetrate beyond the risk of desiccation. In the possession of this extensively ramifying root which enters the soil and absorbs nourishment, the Botrydium differs from algæ in general, which have no genuine root, as they absorb nourishment through the whole of their tissues from the surrounding medium. It is truly a terrestrial plant, and is incapable of being developed under water, for submersion even for a few hours has the effect of bursting its globular head, and thus the spherules which it contains are set free and floated away from the parent to form new colonies. It also develops new individuals from its roots by a process of budding or vegetative increase. The Botrydium is a perfect miniature plant, with root, stem, bud, and fruit, in imitation of the most highly developed flowering plants, but strange to say, it is unicellular, consisting of one continuous' cavity running through the entire plant. There are some very curious and little-known green algæ allied to this plant, which are furnished with

similar adaptations, as their fronds are so in-
crusted with lime as to render nutriment through
the surface precarious. They resemble cacti,
reticulated corals, flabelliform corallines, little
wheels fixed on delicate stems, etc., and are very
beautiful in their shapes or in their structure,
when divested of the carbonaceous coating in which
they are masked. Caulerpa, Halimeda, Acetabul-
aria, etc., are examples of these curious organisms,
which might easily be overlooked as corals. They
are all natives of warm climates, such as New
Zealand and Papua. The *Caulerpa prolifera*, like
the Botrydium, consists of a single cell, though
it is often a foot in length, and is branched with
what has the appearance of leaves and roots.

In pools and ditches of fresh water may often
be seen vast masses of the two-pronged filaments
of *Vaucheria dichotoma*, each filament being sometimes two feet long, almost rivalling the huge
masses of *Cladophora mirabilis*, and the *Conferva
melagonium* of the Arctic regions. It differs little
from the Botrydium except that the spherical
vesicle of the latter is elongated into a simple
or branched thread. Another species is very
common on the ground in damp, dark situations,
such as the ledges and crevices of cliffs in sub-
alpine glens, and is also occasionally observed in
gardens on walls or unfrequented walks, creeping

over the earth in a very thin intricate fleece of a bright grassy-green colour. The filaments are tubes containing an internal green pulverulent mass like the other confervæ ; but the fructification is developed on the outside in the form of dark green homogeneous vesicles attached to the filaments. The Vaucherias are plants of higher organization than most of the other confervæ, and are distinguished by the comparatively enormous cells of which they are composed. They may be known from all other confervæ with which they may be associated by the fact of their branching without joints. They are indeed composed of only one cell, sometimes attaining many inches in length. They are reproduced by very large oval gonidia, which are covered with innumerable cilia, and in consequence endowed with active motion, while they are vivified by the agency of spermatozoa. This mode of reproduction by motile spermatozoa, strange to say, connects the lowest class of plants with the highest members of the animal kingdom in the exercise of their most important function.

The various species of confervæ are known in country places by the popular name of crow-silks, and are used when dried for stuffing beds, for making wadding for garments, and some of them even for manufacturing paper. Pliny mentions that, in his time, they were in much repute as a healing remedy for fractured limbs. They sometimes abound to such an extent as to be positively injurious to health. After floods, for instance, when the overflow stands several days, they grow and spread with such rapidity, as on the subsidence of the water to form a uniform paper-like mass, to which the name of meteoric paper has been given. Till the stratum becomes perfectly dry, which is a slow process, except on the outer surface, the smell is often very disagreeable, and the gas generated from it renders the meadows extremely unwholesome. Every one must have remarked the unpleasant odour exhaled by streamlets when their waters begin to fail in a hot summer, and thus expose the masses of confervæ which they contain. Specimens of the so-called meteoric paper have been preserved in the library of Bernhedin. One side is smooth, and of a brownish-ash colour, the other of a greenish red-brown. One of the pieces preserved was thirty-four feet long and three feet wide. The grey side was the more compact, and much resembled grey blotting-paper. It received its paler hue from the bleaching effect of the sun's rays.

The fresh-water algæ of foreign countries are very similar to our own. Fewer differences exist between native and foreign species of this family, than between those of any other family of plants. Every traveller is struck with the wonderful sameness of fresh-water productions animal and vegetable all over the world. De Candolle has remarked that in large groups of plants that have many terrestrial and only a few aquatic species, the latter have a far wider distribution than the former. Brazil, whose land flora and fauna are quite different from those of Britain, has yet many fresh-water insects, shells and plants precisely similar. Many fresh-water and marsh plants have an immense range over continents, extending even to the remotest islands ; while water-beetles and fresh-water mollusca present the same types all over the globe. This similarity of type arises doubtless from similarity of conditions, and also from the temporary nature of collections of fresh water as compared with land and sea, not giving time sufficient for the production of varieties or of new species, which we have reason to believe is a very slow process, constantly exposing species of restricted range to destruction, and allowing only families of wide distribution to be preserved. As a rule, certain forms of fresh-water algæ are to be found only in certain localities. The most conspicuous species in stagnant waters are Oscillatorias and Zygnemas. Other species, such as *Conferva glomerata*, and

different *Rivularias*, affect running water ; others
are found only in wells, such as the beautiful and
delicate *Draparnaldias;* and others are partial to
lakes, such as the *Chætophoras.* The Irish lake
of Glaslough is remarkable for its greenness
occasioned by *Oscillatoria æruginosa. Ulva ther-
malis* lives in the hot springs of Gastein, the
temperature of which is 117° Fahr. *Calothrix
nivea* luxuriates in the sulphuretted hydrogen
water of Harrogate. *Protococcus salinus* gives
a crimson colour to salt-water tanks on the
coast of the Mediterranean. *Hæmatococcus Noltii*
reddens the marshes of Sleswick-Holstein. Bally-
drain lake in Ireland is coloured a lovely green
by *Anabaina spiralis.* The fruiting season of
the fresh-water algæ is in spring and early
summer. They rise then from the bottom of
the water, where they lay all winter unseen ; and
buoyed up by the globules of gas which they
eliminate, and which gives them a vesicated or
bulbous appearance, they float on the surface in
large masses. When fruiting they lose their
bright green colour and become dingy, often
yellowish and very dirty-looking. Such specimens the tyro is apt to pass by as worthless, but
they are the most valuable in a scientific point
of view ; while the vivid green slimy masses of
barren filaments that excite his admiration as

they lie at the bottom of the ditch or pond are rarely worth carrying home.

But the most extraordinary of all the members of this numerous and highly varied family of plants are the Diatoms or Brittle-worts, which form a wonderful microcosm of their own. It is but a few years, comparatively speaking, since the microscope has drawn aside the veil which hid them from our view; but our knowledge of them, thanks to the all-absorbing attention with which scientific men have regarded them, is already remarkably extensive and accurate. Though these curious vegetable atoms occupy the lowest place in the scale of vegetation, they are, nevertheless, intensely interesting and sugges- tive of marvellous thought. They constitute an immense family, the individuals of which are numerous beyond the sands of the sea-shore or the stars of heaven; ay, even beyond the wildest dreams of the Pantheist. They cannot be reckoned by millions simply, but by hundreds of thousands of millions. There is hardly a spot on the surface of the land, or in the depths of the ocean, where some species or other of them may not be found either in a dead or living state. They inhabit streams, ditches, and stagnant pools; they clothe the leaves and fringe the stalks of sea-weeds; and they are found in inconceivable multitudes amid the mud and detritus deposited by rivers at their mouths, and by the accumulation of their exuviæ, year after year, occasion a vast deal of labour and cost to the dredger. The mud of the Nile and the Ganges, which have formed the great deltas of Egypt and Bengal, is full of them. Naturalists, who have explored the virgin forests of the tropics, inform us that the very branches of the trees are covered with vast numbers of them. They have been discovered in the stomach of the oyster, the clam, and the barnacle ; and Dr. Hooker says, in the *Botany of the Antarctic Voyage*, that the stomachs of the salpæ and other molluscous animals, which were washed up in immense masses on the ice, invariably contained several species of diatoms. On the soil of our fields they occur in myriads among guano, the product of those vermivorous shore-birds which inhabit the desolate islands of the South Seas ; and on the tops of the highest British mountains—Ben Lawers, Ben Nevis, and Ben Macdhui—I have repeatedly gathered them in great quantities from the black mud which is generally found under masses of melting snow. The ice-bound seas of the north are peopled by them. They form the brown staining matter of the 'rotten ice' so well known to all Northern

navigators. Along with various species of animalcules, they are the cause of that peculiar
olive-green tinge which extends over .a portion of
the Arctic ocean, amounting to not less than
20,000 square miles, every two miles of which,
according to Scoresby's estimate, comprehends
23,888,000,000,000,000,—a number which would
have employed 80,000 persons since the creation
to reckon![1] In the Antarctic Ocean, on the other
hand, far beyond the limits where even the hardy
lichen, moss, and sea-weed refuse to vegetate
upon the rocks, and where every circumstance
would seem inimical to the growth and propagation of even the simplest plants, they occur in
countless myriads on the floating ice, and cover
the sea with meadows of a pale-brown hue,
extending as far as the eye can reach, and down
from the surface of the water to abysses deeper
than plummet ever sounded. They form an

[1] Scoresby says, 'After a long run through water of the common
blue colour, the sea became green and less transparent. The
colour was nearly grass-green with a shade of black. Sometimes
the transition between the green and blue water is progressive,
passing through the intermediate shades in the space of ten or
twelve miles; at others it is so sudden that the line of separation is
seen like the ripple of a current, and the two qualities of water
keep apparently as distinct as the waters of a large muddy river on
first entering the sea. In 1817, I fell in with such narrow stripes
of various coloured water, that we passed streams of pale-green,
olive-green, and transparent blue, in the course of ten minutes'
sailing.'

R

enormous bank, flanking at an average depth of 1800 feet the whole length of Victoria Barrier—a glacier of ice some 400 miles long and 120 broad. And it is extremely probable that they are uniformly dispersed over the whole surface of the ocean ; for, owing to their extreme minuteness in their individual state, and the transparency of their tissues, they cannot be perceived by the naked eye unless when accumulated into immense masses and contrasted with opaque substances. The surface of the sea, it has been said, is one wide nursery, its every ripple a cradle, and its bottom one vast cemetery. The floor of the ocean is paved with these organisms; those mysterious submarine plains, where the seer's vision of the 'sea of glass' seems realized, where no wind blows, and no storm rages, and no current frets, are covered with their remains unmixed even with a single particle of sand. The soundings obtained from these silent motion-less depths, are as pure and free from the slightest intermixture of other matter, as the new-fallen snow-flake is from the dust of the earth. And as a snow-cloud in a still January evening discharges its wavering flakes upon the earth, so are the waves continually letting fall upon their bed showers of minute diatoms whose term of life had expired, kindly strewing the melancholy wrecks of ships with their fleecy coverings, and protecting by their soft cushions the floor of the deep from the abrasion of the waters.

Humble and minute although these diatoms may be, they are among the oldest of the living inhabitants of the globe, having performed their part in creation long ages before the first parents of the human race were called into existence. The wonderful records which they have left behind them in our rocks carry us back to a period when the world, now so beautiful with its verdant meadows and waving woods, was one dreary pestiferous bog, where calamites, sigillarias, and other gigantic marsh plants formed intricate jungles, in whose damp recesses horrid reptiles roared and wallowed, and made war upon each other. In the waters of the primeval seas they flourished in the greatest profusion, supplying the ultimate food of the pleiosauri, ichthyosauri, and the other huge reptiles with which they swarmed, just as their successors form the basis of subsistence, through an amazing series of links, for those mighty devourers, the whales, the seals, and the walruses of the Arctic and Antarctic oceans. The fiery cataclysms, which extirpated whole races of plants and animals, left these atomies uninjured ; the physical changes going on over the whole earth only served to carry them uninjured from one geologi-

cal epoch to another, until at length we behold in
the diatoms of our pools, rivers, and seas, the representatives and exact counterparts of the races
that lived and died in those ages of the world,
compared with which the antiquity of recorded
time is but as yesterday. Step by step, up from
the lowest fossiliferous strata, when life was just
feebly dawning, when the eye that gazed upon
the dreary lifeless scenes which the earth then
presented was more rudimentary than that of the
mollusc, and the ear that listened to the wild
ceaseless moaning of waves, the splintering of
rocks, and the roar of volcanoes, was but a mere
otolithic vesicle; through the old red sandstone,
with its numerous strange and monstrous fishes;
the carboniferous strata, with their countless forms
of gigantic vegetable life ; and the limestone rocks,
the graves of whole hecatombs of madrepora,—
through all these different geological deposits we
can trace the presence of these little plants. Endowed with the power of investing themselves, as
if by a mysterious process of electrotype, with
the silicious matter held in solution by the waters
in which they abound, they are in truth indestruc-
tible ; and of their remains, individually so minute
that hundreds may be contained in a drop, and
thousands packed together in a cubic inch, deep
beds of marl, extensive chains of hills, huge lime-

stone rocks, ay, even whole territories of alluvial soil, have been in a great measure composed.

In Virginia there are vast beds of silicious marl, composed of the skeletons of countless generations of diatoms ; and it is said that the towns of Richmond and Petersburg, in the same province, are built upon an enormous stratum of these plants, every cubic foot of which contains billions more than the living population of men that throng the streets above them. Extensive tracts covered with similar relics of a former age occur throughout Britain. The peat mosses of Ireland and the Highlands of Scotland abound with them, and hundreds of species have been found beautifully preserved in the vast amber beds of Prussia. The peculiar white powdery substance known by the name of *Berg mehl,* or mountain meal, found in Swedish Lapland, under beds of decayed moss, and mixed by the inhabitants with their food in times of scarcity, is composed of fossil diatomaceæ, several species of which are still living, and occasionally seen in this country. The *fossil flour* which the Chinese mix with their wheat or rice on similar trying occasions ;[1] the unctuous clay which

[1] The following particulars regarding Chinese fossil flour, adapted from Ehrenberg's late great work, *Mikrogeologie,* may be interesting :—

'Various kinds of edible earth were known in China in very ancient times, and it may be presumed that many of them are mixed or pure tripolitan fresh-water bioliths, *i.e.,* species of earths or

the Otomacs gather on the shores of the Orinoco and
Meta, and eat by way of a *bonne bouche* after their
regular meals ; the yellowish earth called caouac
found in Guinea, of which the negroes are passion-
ately fond ; the kieselguhr or meerschaum, found
in Hamburg and Turkey, and employed in the
manufacture of pipes, and also recently on account
of the extraordinary greed with which it absorbs
and retains nitro-glycerine, as a mechanical medium
or vehicle in the conversion of that explosive
liquid into the safer and more useful form of solid
dynamite ; and the polierschiefer, or polishing
stones, the elements of which consist chiefly of remnants of microscopic living beings.   In the year 1839, Biot read before the
Academy of Sciences in Paris a treatise, containing everything that
was then known on this subject, to which his son, the Oriental
linguist, Biot, furnished translations from Chinese and Japanese
works.   From Schott, in Berlin, Professor Ehrenberg obtained, in
addition, the following information, taken from Chinese sources.
The first mention of edible earth dates from the year 744 after
Christ, and is contained in the Chinese work, Pen-tsao-kang-mu,
where it is called Schimian, Stone-bread, or Mi-anschi, Bread-
stone; the article in the Japanese *Encyclopædia*, which Biot has
translated, is taken from this work.   The Pen-tsao says, according
to Schott, that stones contain several substances which are edible,
especially a yellow meal and fatty liquid, which is contained in the
Yu (a stone), and is therefore called the fat, marrow, or mucilage
of the white Yu.   An earthy substance, prolonging life, and called
Schi-nao, is found in the very smooth stone Hoa-shi, which is supposed to be Steatite, and may perhaps be decomposed Steatite.
The Schimian is only used as a substitute for bread in times of
scarcity, when it is miraculously found in different localities, as is
believed.   The Imperial annals of the Chinese have always re-
ligiously noticed its appearance, but have never given any description of the substance.   The Pen-tsao quotes, under the Emperor

slate of Bilin, which supplies the tripoli used for polishing stones and metals, are all found, when subjected to the microscope, to consist almost entirely of the silicious plates of diatomaceæ, united together without any visible cement. The world, it has been well said, is a vast catacomb of diatoms, a grand herbarium in which these most ancient plants have been preserved in a state of completeness and accuracy little short of their living perfection, to be to us the unimpeachable records of time, as it were, beyond time, of mountains and shores, rivers and seas, that seem mythical even to the geologist.[1] They were at work

Hiuan-Tsung of the great dynasty Tang, in the third year Tainpao (744 after Christ), a spring in Wujin (now Liang-tschen-fu, in the province Kan-su), which ejected stones that could be prepared into bread, and were gathered and consumed by the poor. (Schott.)

' Under the Emperor Hian-Tsung of the same dynasty, in the ninth year of the period Yuen-ho (809 after Christ), the stones became soft and turned into bread. (Biot.)

' Under the Emperor Tschin-Tsung, of the dynasty Sung, in the fifth year of the period Ta-tschong-Tsiang-fu (1012 after Christ), in the fourth month, there was a famine in Tsy-tschen (now Ki-tschen in Ping-yang-fu, in the province Schan-si), when the mountains of Hiang-ning, a district of the third rank in the same part, produced a mineral fat (Stone-fat) resembling a dough, of which cakes could be made. (Schott.)

' Under Jin-Tsung, in the seventh year of the period Kia-yeu (1062), stone meal was found.

' Under Tschi-Tsung, in the third year of the period Yuen-fong (1080), the stones turned into meal. All these kinds of stone-meal were collected and consumed by the poor. (Biot.)'

[1] As the earliest fossil diatoms yet found, judging from the figures of Ehrenberg, are identical in every point with the great

in the primeval world long before man was ushered upon the scene, and they are at the present day employed in altering and modifying the grand features of the globe ; in producing results which man is as incapable to predict as he is powerless to prevent. Who is there that can gaze upon these wonderful plants, which thus, as it were, connect the ages and the zones, without a dizzy sense of the infinity and permanence of nature, and the power of Him whose judgments are unsearchable, and whose ways are past finding out ?

majority of species now living in our waters, and forming deposits which will become rock at some future time ; and as some species are peculiar to lakes and rivers, and others to seas and firths, while some affect deep and others shallow water, these tiny plants are capable of furnishing considerable information to the geologist, with regard to the conditions under which raised sea-beaches and fresh-water limestone rocks were originally deposited, and the circumstances which operated in the production of the different strata in which they occur. I may add, as an illustration of the universal diffusion of these plants, the curious fact, that the late Dr. Gregory found numerous most interesting diatomaceous forms in small fragments of soil not exceeding a pinch of snuff, adhering to specimens of exotic plants in herbaria. In every case, without exception, he found these organisms ; and in all, the proportion to the whole non-calcareous earthy residue was wonderfully large. The soils in which the most numerous species were found, were respectively obtained from the Sandwich Islands and Lebanon. Many of Ehrenberg's profound observations were made on portions of foreign soil procured in this manner, and his example should stimulate collectors of plants to preserve carefully every vestige of earth adhering to the roots of their exotic specimens, as in this way many new forms may be brought to light, and many rare ones studied in the quiet and leisure of home, without the trouble and fatigue of collecting them in their native localities.

But this is not all! Wonderful as it may seem, the very realms of the air are peopled with diatoms. The atmosphere we breathe contains hundreds of species, which float about on every breeze, and are wafted hither and thither. Many of them remain for years in the highest strata of the atmosphere, until carried down in the full capacity of life to the nourishing waters of the stream and the lake, by descending currents of air. They have been found in immense numbers in the impalpably fine dust, which at certain seasons broods like a thick haze over the island of St. Domingo, and occasionally falls in great quantities on the decks of vessels far out on the Atlantic. The sirocco and trade winds convey immense quantities of them for hundreds of miles. Clouds of diatomaceous dust, giving the atmosphere an orange or ochre hue, have repeatedly been observed coming in various directions from the coast of Africa, falling on vessels, and diffusing around a darkness so dense as often to cause them to run ashore. Similar showers are not unfrequent in China, and spread over several provinces at once and far out to sea. They are raised from the Mongolian steppes—regions of sand more than 2000 miles long and 400 broad—and falling into the waters of the Yellow Sea, give it that peculiar tinge from which it derives its name. During the

dry season in the lifeless plains of the Orinoco,
and the great Amazonian basin, when the soul is
parched and triturated by the intense, long-con-
tinued drought, dense clouds of diatomaceous dust
are raised by the winds and wafted to great distances. These showers happen most frequently
in spring and autumn after the equinoxes, but at
intervals varying from thirty to fifty days. From
the nature of the species wafted by these winds,
the region which originally produced them can be
ascertained with tolerable accuracy ; and hence
they afford a clue to those mysteriously wayward
aërial currents, and cyclical relations in the upper
and lower atmosphere, which have hitherto per-
plexed meteorologists. It has been observed that
these storms in certain districts, amply compensate
for the annoyance they occasion. The soil of the
countries most subject to the visitation, when of a
compact character, is loosened and lightened by
the dust, and at the same time the lighter fertilizing matters carried away by the great rivers are
replaced by organic remains, so that an abundant
harvest follows the devastations committed by
these dust showers. Nearer home these curious
meteoric phenomena have occasionally been observed. Black rain, composed of· portions of decayed plants, mixed with the skeletons of diatoms,
fell in Ireland in April 1849, over a district of 700

square miles. A great mass of substance remarkably like paper fell during a violent storm in 1687,
near the village of Randen in Courland, which
excited great curiosity at the time, and was found
after the lapse of many years, by the all penetrating microscope of Ehrenberg, to consist of a com-
pactly matted heap of diatoms and confervæ.
Diatoms have even been discovered in the pumice
and ashes ejected from the burning craters of
volcanoes.

'The dust we tread upon was once alive!' was
the exclamation of one great poet ; and 'How
populous, how vital is the grave!' was that of another, but little did either Byron or Young know
how extensively true were the words they uttered.
The microscope shows us how inconceivably populous is the whole world, when thus the loftiest
regions of the atmosphere, and the fathomless
depths of the ocean, and the darkest, deepest
abysses of the earth, where we should suppose all
life impossible, are peopled with myriads upon
myriads which the Infinite mind alone can enumerate, of minute vegetable organisms, performing their allotted task in the great workshop of
nature, and adding a thousand times more to the
mass of materials which compose the crust of the
globe, than the bones of elephants and whales.

To the investigation of the diatoms, we must

not bring any of our preconceived notions of vegetable forms and structures, for we shall assuredly find them completely overthrown, by the new and strange modes of organization which these minute plants display. Indeed, so peculiar and abnormal are some of these modes, so unlike those of all other plants, that the zoologist and botanist are not yet fully agreed as to which kingdom of nature— the animal or the vegetable—they ought to be referred ; and, accordingly, they have occasionally been classed and figured as plants by one naturalist, and as animals by another. Ehrenberg, the great Prussian naturalist, whose microscopic researches have laid open to us a new and strange world of minute organic existence, and to whose untiring industry and patience we are indebted for the discovery of most of the wonderful atomies under consideration, was from the very first firmly convinced of their animal nature ; and the credit attached in this country to his notions, had the effect of turning away the attention of botanists from them ; while the zoologists rejected them from their systems as suspicious and anomalous objects ; and the mere microscopist regarded them simply as new and strange forms of life, with the contemplation of whose beautiful structure he could agreeably while away a leisure hour. In external form the diatoms present remarkable

similarities to many species of infusorial animalcules, and exhibit the same spontaneous movements; and even in their elementary composition they are identical with some of the lowest
members of the animal kingdom. In these primitive plants and animals, we may fairly enough
conclude that the animal and vegetable kingdoms
pass into each other; they form the one common
base or point from which these two systems of life
start, to recede so widely from each other in the
large and complicated organizations which stand
at the head of both. 'From man to the primary
animal and vegetable cell,' Schmidt justly observes, 'there exists no gap in the realization of a
general idea upon which nature as a whole is
based. There is no abrupt transition from one
kingdom to another, but an insensible gradation.
Thus the embryo germ of an alga or sea-weed is
identical, in elementary composition and form,
with that of a medusa or ascidia ; in the former we
have the higher stage of development of the plant,
in the latter the simpler form of the animal.' The
vegetable nature of the Diatoms is, however, I
think, clearly indicated by the marked results of
the application of the spectroscope to them. The
spectrum of *diatomin* or the olive-yellow endochrome of diatoms is absolutely identical with that
of *chlorophyll* or the green endochrome of plants.

The spectrum in question is a very characteristic
one, and cannot be mistaken. It exhibits a very
black, narrowish band in the extreme red, reading at
the lower edge, which appears to be remarkably
constant, about $\frac{7}{8}$ of Mr. Sorby's scale.

The forms which the diatomaceæ assume are
exceedingly varied and beautiful. Most of them,
as already mentioned, are invested with a very
thin transparent glass-like pellicle, engraved with
median lines and transverse striæ, the patterns of
which are wonderfully constant in the same species,
and afford admirable tests for the general excel-
lence of the object-glass of the microscope ; the
distance between the different markings being
often the $\frac{1}{30000}$th part of an inch, and some, it is
even said, being only the $\frac{1}{130000}$th of an inch separate, requiring for their distinct determination a
magnifying power of twelve hundred diameters,
and the aid of oblique light. What appear as
striæ with low powers, assume the form of moniliform or pearl necklace-like markings when examined with high powers. Their silicious investment
has cellulose for its base. The silex is infiltrated
to a variable extent in the different families, and
the mode of its deposition can to a certain extent
be ascertained by examination with polarized light.
In order that the striæ or markings may be clearly
seen it is necessary that the valves or frustules

should be boiled with strong nitric acid, and carefully and repeatedly washed. Some species consist of chains of parallelograms (Fig. 24), connected together at one single point, more beautiful in ap-

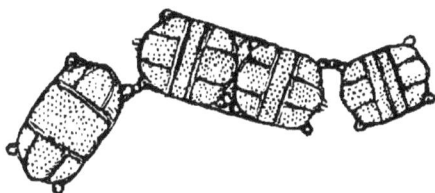

FIG. 24.—DIATOMA BIDDULPHIANA—magnified.

pearance, and more richly and elaborately carved than the costliest bracelet on the arm of a queen.

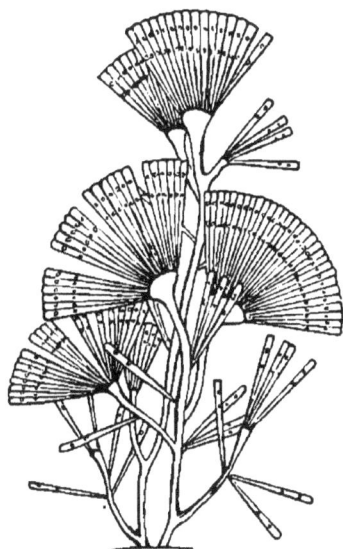

FIG 25.—EXILARIA FLABELLATA.

Some resemble miniature flags or fans, adorned with the most exquisite figures; some graceful boats, frosted and granulated, in which a tiny animalcule might float over a dew-drop; and some little trees (Fig. 25), covered with variegated leaves, arranged in fanlike clusters, as though intended for microscopic models of a grove of fanpalms. In short, they form circles, triangles, squares, and almost every kind of mathematical figure (Fig. 26), to the utter subversion of all the ideas of vegetable forms which we are accustomed to entertain.[1] Diatoms are generally colourless; but some species are of a deep green, or rich brown, or a pale yellow or red. They are delicate as hoar-frost, and seem more like the strange vegetation produced on our window-panes on a cold frosty morning, than veritable living plants. These little organisms, we must not forget, exquisitely beautiful and curious in form and structure as we find them under the microscope, ap-

FIG. 26.—ACHNANTHES UNIPUNCTATA—both figures magnified.

pear to the naked eye a mere green or dark-brown film, or indefinite slimy scum, on the leaves of an aquatic moss, or the stalk of a sea-weed !

The propagation of the diatomaceæ is performed in a very simple manner. At certain stages of

[1] The deposits from Franzenbad, San Fiore in Tuscany, Bilin, Bermuda, and Lough Morne in Ireland, are well known as containing many of the most beautiful species, and specimens of them are sold by dealers in microscopic objects and apparatus. Most of the curious forms that are unknown in this country may be obtained from Peruvian guano.

their growth, the frustules, or fragments of which they are composed, separate in some species into two portions, each of which forms around itself a cell-wall, possessing a form and character precisely similar to those of the original one ; and thus a very material increase in the number of frustules is, through course of time, effected. This process is called fissiparous or merismatic division. It is nothing more than what Professor Huxley calls 'a process of discontinuous growth.' What in the higher plants is a process of growth, an enlargement and development of the individual, is in these simple organisms a process of multiplication of separate individuals. In some cases the process of reproduction is performed by the conjugation of two approximated frustules, as was seen in the case of the larger confervæ, the result being the union of their contents by means of interposed tubes, and the subsequent production of a germinating spore ;—thus leaving their vegetable origin no longer a doubtful question. Professor Weiss regards the large cavity between the two frustules as analogous to the embryo sac of higher plants ; and he has succeeded in observing the development of new individuals in it. The product of this new individual indicates the alternation of generations in the Diatoms. By these various methods they propagate themselves with incon-

S

ceivable rapidity ; and hence it is not difficult to account for their almost universal diffusion, and the enormous accumulation of strata which they form in certain places.

Closely connected with the Diatoms are the *Desmidias*, which have attracted almost as large a share of attention among microscopic observers. They are equally remarkable for their universal presence and abundance, and for the variety and beauty of their forms. Microscopic in their individual state, when adhering in large quantities to Potamogetons and other aquatic plants, or lying at rest at the bottom of pools, they form a green perceptible film or coating. Unlike Diatoms, which are found in salt as well as fresh water, the Desmids are exclusively denizens of fresh water, preferring that which is pure and limpid, and always most abundant not in shady places or woods, but in open pools in exposed situations. Owing to their isolated, unattached condition, they love to dwell in quiet shallow waters, never growing in rapidly running or very deep water ; the larger species being generally found nearest the bottom. Under the microscope they are remarkable for their singular shapes and their external markings and appendages. As a rule they are devoid of the silicious envelope which characterizes the Diatoms, and their markings are always elevations and not depressions as in these plants. They are usually of a very deep vivid green colour, owing to the great quantity of chlorophyll which they contain. Individually their forms are oval, crescent-like, cylindrical, oblong, or with variously shaped rays or lobes giving them more or less a stellate appearance ; but in their social state, when aggregated together, they are often arranged in linear series, collected into faggot-like bundles, or in elegant star-like groups, enclosed in a common, very transparent gelatinous coat. As a class they are characterized by their bilateral symmetry. Each frustule is in reality a single cell, which has a tendency to divide into two valves or segments ; and this tendency is indicated by a constriction which is more or less deep in different genera. Owing to their power of locomotion, Ehrenberg and his followers regarded Desmids as of animal origin ; but this property, as we have seen, belongs to a large number of the lowest vegetable organisms. Their plant nature is now conclusively established. When mixed with mud, they are able to make their way to the surface ; and when contained in a bottle they begin to congregate on the side that is most exposed to the light. In some instances they have even been known to retire beneath the surface of the mud of ponds before it dries up, after the water has been drained off. But not only does the whole Desmid move itself in the water in this way, its cell-contents also exhibit various movements similar to those of Chara, Vallisneria, and Anacharis. Large globular granules flow in uninterrupted currents on the inner surface of the utricle. The course of the currents is not very determinate, and they seem to pass each other in close proximity, continuing however for hours moving in the same manner. By using the fine adjustment, a single granule may often be followed in its course round the end of the cell, down the edge and across the suture, thus affording a beautiful demonstration of the unicellular character of the plant. Like many other fresh-water algæ, the Desmids are ordinarily reproduced in two ways, by simple cell-division, when each frustule divides into two; and by the conjugation of the two cells of a single filament, or of two separate filaments, producing by their organic union and the blending of their endochrome, a spore whose granular contents become gradually brown and red, while their coats become thickened, granular, or even spinous, resembling bodies found fossil in flint, supposed to be the spores of Desmids.

But before concluding this chapter I must refer to an organism which has for a long time been considered by all authors to be an animal, but is now conclusively proved to be a fresh-water alga. This is the well-known *Volvox globator*, which, like the *Protococcus nivalis*, consists of a single cell, possessing the power both of nutrition and reproduction. It is a rolling crystal sphere studded with emeralds ; or to use less poetical words, it is a symmetrically formed sphere composed of perfectly colourless transparent membrane, with colourless watery contents, without any aperture, and studded over at equal distances with small green spots, in quantity ranging from 30 to 300. These green spots are identical with the ciliated zoospores of other algæ. Owing to the associated movements of these zoospores, their cilia projecting through the enclosing membrane, the whole full-grown plant moves freely in the water with a graceful motion, sometimes gliding slowly across the field of the microscope, then stopping and revolving, and then combining both rotation and progression. It reproduces itself in a remarkable manner, no less than three generations being present within the parent envelope at one time ; daughter-cells springing from the zoospores, and known by their greener colour, and grand-daughter cells produced from the daughter-cells by segmentation. This exquisite organism is found in most open clear ponds, whose water is free from sewage and maintained at a uniform level all the year

round. It is most abundant on the sunny margins
during the spring and early summer months.
Gazing upon these ' moving spheres ' in the water,
we are profoundly impressed with the thought that
motion is everywhere—above us, around us, beneath us. Still and fixed as the stars appear, they
are revolving in space with inconceivable velocity,
and are the centres of forces and movements of
the most stupendous nature. Were our vision endowed with more than telescopic power, and were
these sublime motions, separated by vast intervals
of time, compressed into a day or an hour, we
should find that everything like rest in special existence had forthwith disappeared. We should be
overwhelmed with the roar and the speed of the
blazing spheres. Similarly the stars of the earth
beneath our feet which revolve in the orbits of
the seasons, seem motionless and spell-bound in a
magic stillness. And yet, quiescent as they appear,
could we penetrate with more than microscopic
vision beneath their calm exterior, we should see
miniature worlds of amazing activity, that would
bewilder and distract us. But this motion is so
frozen by distance and minuteness that we live
between two great worlds of silence, the silence
above us in the stars of heaven, and the silence
beneath us in the grass of the fields, and we feel
nature only as a soothing and solemn rest.

Such is a brief and imperfect sketch of the history and peculiarities of these wonderful freshwater algæ. They open up to us the infinitude of microscopic life, reveal a vast and glorious realm of new creative design, whose limits can never be fathomed, and whose mysteries can never be exhausted by man's finite researches. It is not so much what they actually disclose that awes and astonishes us ; but the bewildering boundlessness of the unknown arcana beyond, to which they point. The vast additions which they have made to our knowledge, have only left the immensity of the universe of life greater and more mysterious than before. For it is all but certain, that if our vision could be made more piercing, and our instruments more perfect, while we explored onwards through the successive realms of the invisible towards the inmost shrine of nature, we should find new scenes of wonder and beauty continually unfolding themselves, and new fields of omniscient display constantly revealing to us that God was still before us in all His exhaustless, creative energy, and that we saw but the 'hidings of His power.'

CHAPTER IV.

ATURE is a perpetually revolving pano-
rama. No sooner does she withdraw one
object from our admiring gaze, than she
immediately places another as interesting or as
beautiful in its room. In watching the progress
of vegetation especially, as month after month it
unfolds before us, we are struck with the regularity
with which each species of plant visits us in its own
appointed time. So remarkably constant are the
same plants to their appointed seasons, that their
appearance might be regarded as a kind of floral
calendar, indicating the various periods of the
year. This regularity is not confined to the
highest tribes of plants, but is equally observable
in the very humblest. The smallest and most ob-

scure tribes have some peculiar functions adapted to each period of the year. Though most of them are perennial, yet they are more luxuriant in some seasons than in others, and are particularly exact and exclusive as to their periods of reproduction. The hard and apparently lifeless lichen remains unchanged upon the rock for years, perhaps as long as the rock itself continues uncrumbled, but every year at the approach of winter, when the moist, stormy weather in which it delights prevails, its dormant suspended life revives, and when higher plants are hybernating, it begins to exercise the various functions of vitality. The bright silken tufts of the moss continue throughout the whole year to soften the rough harsh aspect of the wall and ruin, and to form velvet pads on the woodland walks to hush the fall of fairy feet, but in spring when 'a fuller crimson comes upon the robin's breast, and a young man's fancy lightly turns to thoughts of love,' it awakens under the ethereal influence of the universal feeling, clothes itself in its fairest robes, and puts forth its crimson urns, that burn like fairy love-jewels among its emerald leaves. The naiad-like confervæ vanish from the waters, for nine months in the year, and return to luxuriate in their cool, clear haunts, as duly as the warm breath of April melts away the icy fetters from the rejoicing streams, and once more,

> ' Inverted in the tide,
> Stand the grey rocks, and trembling shadows throw,
> And the fair trees look over side by side,
> And see themselves below.'

While the approach of autumn is unmistakably indicated by the springing up of mushrooms in the moist dark recesses of the woods, even when the viewless boundary of summer is not yet crossed, and the air is still balmy and sunny, and the robe of nature fadelessly green.

Fungi are intimately associated with autumn ; unrobed prophets that see no sad visions themselves, but that bring to us thoughts of change and decay. Indeed, so close is this association that they may be called autumn's peculiar plants. The blue-bell still lingers on the wayside bank, and in the woods a few bright but evanescent and scentless flowers appear, but fungi and fruits form the wreath that encircles the sober and melancholy brow of autumn : fruits the death of flower-life ; fungi the resurrection of plant-death. The seasonal conditions which arrest the further progress of all other vegetation, which cause the leaf to fall, and the flower to wither, and the robe of nature everywhere to change and fade, give birth to new forms of plant-life which flourish amid decay and death. From the relics of the former creations of spring and summer reduced to chaos, springs up a new creation of organic life ; and thus nature is not a mere continuous cycle of birth, maturity, and decay, but rather a constant appearance of old elements in new forms.

This new tribe of plants comes in at a peculiarly seasonable time, when the more aristocratic members of the vegetable kingdom have departed, leaving the favourite haunts of the botanist bare and destitute of interest. The collection of them in the field, and the study of their peculiarities in the closet, will furnish ample occupation of a most fascinating nature during the whole season, as new facts always connect themselves with new forms. To those who enjoy mysteries and paradoxes there can be no lack of such enjoyment among the fungi. In many respects they are the most mysterious and paradoxical of all plants. In their origin, their shapes, their composition, their rapidity of growth, the brevity of their existence, their modes of reproduction, their inconceivable number and apparent ubiquity, they are widely different from every other kind of vegetation with which we are acquainted. In studying their history we walk amid surprises ; and as we lift each corner of the veil, more and more marvellous are the vistas that reveal themselves.

The first thing that suggests remark in regard

to these curious organisms, is their origin. Incapable of deriving the elements of growth from the crude unorganized crust of the earth, they are parasitical upon organic bodies, and are sustained by animal and vegetable substances in a state of decomposition. That living and often nutritious objects should spring from festering masses of corruption and decay ; that plants, endowed with all the organs and capacities of life, should start into existence from the dead tree that crumbles into dust at the slightest touch, or draw their nourishment from dried and exhausted animal excretions, which have lain for months under the influence of drenching rains and scorching sunbeams, is indeed a profound mystery of nature. No sooner does the majestic oak yield to the universal law of death, than several minute existences, which had been previously bound up and hid within its own, reveal themselves, seize upon the body with their tiny fangs, fatten and revel upon its decaying tissues, and in a short space of time reduce the patriarch and pride of the forest, which had braved the storms of a thousand years, into a hideous mass of touchwood, or into a heap of black dust. How strikingly do these plants illustrate the great fact, that in nature nothing perishes; that in the wonderful metamorphoses continually going on in the universe there is change, but not

loss ; that there is no such thing as death, the ex-
tinction of one form of existence being only the
birth of another, each grave being a cradle.

In the previous chapters I have incidentally alluded to the relation of fungi to algæ and lichens.
Mr. Sorby has shown by his chromatological researches that their most common colouring matters
exactly correspond with those found in the apothecia of lichens, and also that their more
accidental constituents are quite analogous to
those occasionally found in the fructification of
particular lichens such as the cup-mosses. He
considers that they bear something like the same
relation to lichens that the petals of a leafless parasitic plant would bear to the foliage of one of a
normal character ; that is to say, that they are the
coloured organs of reproduction of parasitic plants
of a type closely approaching that of lichens. In
appearance and mode of decay fungi resemble the
curious parasitic Rhizanths,—a low class of flowering plants intermediate between Thallogens and
Endogens, of which the *Rafflesia Arnoldii* of Java is
an extraordinary example, composed chiefly of cellular tissue, whose seeds closely resemble spores,
and which are never green, but assume a brown,
yellow, or purple colour. The *Cynomonium cocci-
neum* of Malta, long celebrated for averting hæmor-
rhage, was called *Fungus Melitensis* from this

likeness. Fungi may be said to resemble the earliest stage of all flowering plants. During the germination of the seed a close analogy exists between the growing embryo and fungi. Both are supplied with nutriment previously organized, the one by its parent, the other by the decay of animal or vegetable matter ; both are developed most rapidly when supplied with warmth and moisture, and in the absence of light, and both liberate carbon to a large amount without assimilating any from the atmosphere. Indeed, the larger portion of every flowering plant which consists of simple sacs of cellulose, secluded from the light, absorbing organic compounds, and developing no special colouring matter, may be compared to fungi ; the green parts alone performing the great function of the vegetable kingdom in keeping up the balance of organic nature.

In many of their properties, the fungi are closely allied to some members of the animal kingdom. They resemble the flesh of animals, in containing a large proportion of albuminous proximate principles ; and produce in larger quantity than all other plants azote or nitrogen, formerly regarded as one of the principal marks of distinction between plants and animals. This element reveals itself by the strong cadaverous smell, which most of them give out in decaying, and also by the

savoury meat-like taste which others of them afford.
Of all known bodies nitrogen is the most unstable.
Its compounds are decomposed by slight causes ;
and therefore its presence in the animal frame is
the cause of its activity and proneness to change.
To this circumstance also is owing the fugacious
character of fungi, their speedy growth and decay.
Unlike other vegetables, fungi possess the remarkable property of exhaling hydrogen gas ; and the
great majority of species, like animals, absorb
oxygen from the atmosphere, and disengage in
return from their surface a large quantity of carbonic acid. By chemical analysis, they are found
to contain besides sugar, gum, and resin, a yellow
spirit like hartshorn, a yellow empyreumatic oil,
and a dry, volatile, crystalline salt, so that their
nature is eminently alkaline, like animal substances extremely prone to corruption. The
cream-like substance of which the family of Myxo-
gastres is composed resembles sarcode, and exhibits Amoeba-like movements. Some of them
contain such a quantity of carbonate of lime, that
a strong effervescence takes place on the application of sulphuric acid. Fungi feed like animals
upon organic compounds elaborated by other
plants. They contribute in no way as vegetables
to the balance of organic nature.

Another property they possess, which connects

them with animals, is their luminosity. This quality
is very rare among plants, and is almost peculiar
to the lowest orders of animals, particularly those
which inhabit the ocean. A species of mushroom
(*Agaricus olearius*) grows on the olive tree, which is
often luminous at night, and resembles the faint,
lambent flickering light emitted by the scales of
fish and sea-animals kept in a dark place. Anomalous conditions of various species of Polyporus,
Hypoxylon, etc., formerly referred to the genus
Rhizomorpha, from their root-like appearance,
cover the walls of dark mines with long, black,
branchy, flat fibres, and give out a remarkably
vivid phosphorescent light, almost dazzling the eye
of the spectator. In the coal-mines near Dresden,
these fungoid bodies are said to cover the roof,
walls, and pillars, with an interlacing net-work of
beautiful flickering light, like brilliant gems in
moonlight, giving the coal-mine the appearance of
an enchanted palace on a festival night. Mr.
Gardner, in his interesting travels in Brazil, gives
the following account of a remarkable phenomenon of this nature :—'One dark night, about the
beginning of December, while passing along the
streets of the Villa de Natividade, I observed some
boys amusing themselves with some luminous object, which I at first supposed to be a kind of large
fire-fly ; but on making inquiry, I found it to be

a beautiful phosphorescent fungus, belonging to
the genus *Agaricus*, and was told that it grew
abundantly in the neighbourhood on the decaying
leaves of a dwarf palm. Next day, I obtained a
great many specimens, and found them to vary
from one to two and a half inches across. The
whole plant gives out at night a bright phosphorescent light, of a pale greenish hue, similar to that
emitted by the larger fire-flies, or by those curious soft-bodied, marine animals, the Pyrosomæ.
From this circumstance, and from growing on a
palm, it is called by the inhabitants, " Flor-de-
Coco." The light given out by a few of these
fungi in a dark room was sufficient to read by. I
was not aware at the time I discovered this fungus, that any other species of the same genus exhibited a similar phenomenon ; such, however, is
the case in the *Agaricus olearius* (mentioned above)
of Decandolle ; and Mr. Drummond, of Swan
River colony in Australia, has given an account of
a very large phosphorescent species occasionally
found there.' In this country there is a very common fungus called *Corticium cæruleum* growing in
thin bright-blue effused patches upon dead wood,
rails, etc., which is luminous in the dark. And I
remember on one occasion late at night after a
severe thunder-storm seeing about half-a-mile of
the road between Kenmore and Aberfeldy wrapt

T

in sheets of lambent flame, presenting a most extraordinary appearance, through which I had the
greatest difficulty in persuading my horse to pass.
This, the grandest display of phosphorescence I
have ever seen on land, was produced by countless
fragments of rotten wood, threaded with the white
spawn of fungi, that had been swept down upon
the road from the wood above by the heavy rains.
Superstition and ignorance have magnified this
simple appearance of nature into a supernatural
manifestation ; the *ignis fatuus* occasionally seen in
damp old woods, and regarded by the credulous as
a sign of approaching death and an omen of evil,
being often nothing else than the flickering phosphorescence of fungi in a state of decay. It may
be remarked in connexion with this luminous property, that many fungi are capable of generating
considerable heat. Dutrochet ascertained that the
highest temperature produced by any plant, with
the exception of the curious cuckoo-pint of our
woods, was generated by a species of toadstool
called *Boletus æneus.* Such being the curious properties exhibited by these plants, it is not surprising that at one period they should have been regarded as animal productions, formed by insects
for their habitations, somewhat like the coral-
structures of zoophytes and sponges. Though
this view has long been discarded, yet fungi, as

already pointed out, are evidently one of the
links in the chain of nature which unite the
vegetable to the animal kingdom. The analogy
between the higher fungi and jelly-fishes has long
been noticed. The same circular configuration
exists in both. Every one who sees an opened
Geaster or a starry puff-ball is irresistibly re-
minded of a star-fish stranded on the shore.
And the beautiful white laminated coral is called
Fungia agariciformis, from its resemblance to a
petrified mushroom.

Fungi, unlike most plants, are to a great extent
insensible to the influence of light. They commonly prefer damp, close, ill-ventilated places,
where the light, if any, is of a pale, cold, and
sickly character. Within the sheltering darkness
of dense leafy woods—

> ' Some lone Egerian grove,
> Where sacred and o'ergreeting branches shed
> Perpetual eve, and all the cheated hours sing vespers—'

they are to be found crowding together, and are
only accidentally found elsewhere. This propensity to avoid the exposed glare of sunlight, and to
grow in the darkest shade, seems very paradoxical,
when we consider the essential importance of light
among the vital agencies. Even the humblest
lichen, moss, or conferva, will not develop itself in
the same degree of darkness which is essential to
the wellbeing of the fungus. All other plants are

absolutely dependent upon light for their very
existence.   Roses, tulips, sun-flowers, wait upon
the beams of the sun, and live only in his smiles.
They may be supplied with the requisite conditions of heat, air, and moisture, but without light
they will wither and die ; or, if they do seem to
grow, it is only a false, unnatural, and sickly
growth, losing their substance instead of increasing it, and weighing less when dried than the dry
seed from which this amorphous growth proceeded.
Light is not required for the germination of seeds ;
but if the plant be suffered to grow up in darkness,
it merely uses up the store of food contained in
the seed, and when that is exhausted its further
growth is stopped.   But to the influence of light
the fungi are to a great extent insensible.   They
do not disturb themselves or deign to turn towards the light at all ; they continue to shoot out
perpendicularly, horizontally, or even reversed,
just as the surface from whence they spring happens to be directed.   The Geranium growing in
the cottage window, yearningly stretches out its
tender leaves and blossoms to the smiling sunshine without ; and the pea or potato sprouting in
a cellar, which has but one north window, half-
closed, spreads its cadaverous, blanched, and
brittle shoots in the direction of that feeble flicker
of light ; but the fungus points its stalk and its

seed-vessel as readily from as to the light, as un-
consciously downwards to the earth, as upwards
from it. Give it air, warmth, moisture, and undisturbed quiet, and it can live and luxuriate without light. Fungi growing in mines exhibit the
same characteristic colours which they display on
the surface of the ground. Sometimes, however,
species that grow in caves, or in hollow trees,
assume the most curious abnormal forms, their
metamorphosis remaining incomplete, so that instead of producing fructification the whole fungus
becomes a monstrous modification of the mycelium. But whether these abnormalities are caused
by the want of light or by the equable conditions
of warmth and moisture developing the vegetative
at the expense of the reproductive system, as is
the case in flowering plants, is open to doubt. At
all events, their love of seclusion and darkness
gives an etiolated, sickly complexion to the whole
tribe. In consequence of this habit, they are as a
rule the most sombre of all plants, although instances occur in which the prevailing neutral tints
are exchanged for the most brilliant scarlets and
yellows. Green, which is the most frequent of all
colours, the household dress of our mother earth,
more characteristic of ferns, mosses, lichens, and
algæ than of the higher plants, is almost unknown
in the fungi ; and even when it occurs, it is always

more or less of a verdigris tint, and does not appear to be owing to the action of light and oxygen
upon the contents of the cells.

Another of the remarkable peculiarities of the
fungi is the extreme rapidity of their growth, a
peculiarity more frequently to be seen among the
lowest forms of animal life than among plants,
They seem special miracles of nature, rising from
the ground, or from the decaying trunk of the tree,
full-formed and complete in all their parts in a
single night, like Minerva from the head of Jupiter,
or the armed soldiers from the dragon's teeth of
Cadmus, sown in the furrows of Colchis. It has
long been known that the growth of fungi takes
place with great rapidity during thundery weather,
owing, in all probability, to the nitrogenized products of the rain which then falls. One is surprised after a thunderstorm in the beginning of
August, or a day of warm, moist, misty weather,
such as often occurs in September, to see in the
woods thick clusters of these plants, which had
sprung into existence in the short space of twenty-
four hours, covering almost every decayed stump
and rotten tree. In tropical countries, stimulated
by the intense heat and light, the rapidity of
vegetable growth is truly astonishing; the stout
woody stem of the bamboo-cane, for instance,
shooting up in the dense jungles of India at the

rate of an inch per hour. In the Polynesian
Islands, so favourable to vegetable life are the
climate and soil, that turnip, radish, and mustard-
seed when sown show their cotyledon leaves in
twenty-four hours ; melons, cucumbers, and pump-
kins spring up in three days, and peas and beans
in four. But swift as is this development of vegetation in highly favourable circumstances, the
rapidity of fungoid growth, under ordinary conditions, is still more astonishing. These plants
usually form at the rate of twenty thousand new
cells every minute. The giant puff-ball (*Lycoperdon giganteum*), occasionally to be seen in fields
and plantations, increases from the size of a pea
to that of a melon in a single night ; while the
common stinkhorn (*Phallus impudicus*) has been
observed to attain a height of four or five inches
in as many hours. Mr. Ward, in his work *On the
Growth of Plants in closely-glazed Cases*, says of
it : ' I had been struck with the published accounts
of the extraordinary growth of *Phallus impudicus*.
I therefore procured three or four specimens in an
undeveloped state, and placed them in a small
glazed case. All but one grew during my temporary absence from home. I was determined
not to lose sight of the last specimen ; and observing one evening that there was a small rent in the
volva, indicating the approaching development of

the plant, I watched it all night, and at eight in
the morning the summit of the pileus began to
push through the jelly-like matter with which it
was surrounded.   In the course of twenty-five
minutes it shot up three inches, and attained its
full elevation of four inches in one hour and a half.
Marvellous are the accounts of the rapid growth
of cells in the fungi ; but, in the above instance, it
cannot for a moment be imagined that there was
any increase in the number of cells, but merely an
elongation of the erectile tissue of the plant.'   The
force developed by this rapid growth and increase
of the cells of fungi is truly astonishing.   Mon-
sieur Bulliard relates that, on placing a fungus
within a glass vessel, the plant expanded so
rapidly that it shivered the glass to pieces, with
an explosive detonation as loud as that of a pistol ;
while Dr. Carpenter, in his *Elements of Physiology*,
mentions that 'in the neighbourhood of Basing-
stoke, a paving-stone, measuring twenty-one inches
square, and weighing eighty-three pounds, was
completely raised an inch and a half out of its bed
by a mass of toad-stools, of from six to seven
inches in diameter ; nearly the whole pavement of
the town being heaved up by the same cause.'
Every one has heard of the portentous growth of
fungi in a gentleman's cellar, produced by the decomposing contents of a wine cask, which, being

too sweet for immediate use, was allowed to stand unmolested for several years. The door in this case was blocked up and barricaded by the monstrous growth; and when forcible entrance was obtained, the whole cellar was found completely filled; the cask which had caused the vegetable revel, drained of its contents, being triumphantly elevated to the roof, as it were, upon the shoulders of the bacchanalian fungi.

Rapidity of growth in fungi is necessarily followed by rapidity of decay. Though some of the larger and more corky species last throughout the summer, autumn, and winter, and a few are perennial, growing on the same trunk for many years, slowly and almost insensibly adding layer to layer, and attaining an enormous size, yet the vast generality of fungi are very fugacious. They are the ephemera of the vegetable kingdom. The entire life of most of the species ranges from four days to a fortnight or a month; while there are numerous microscopic species of the mould family whose lives are so brief and evanescent as scarcely to allow sufficient time to make drawings of their forms. What a contrast there is between the minute Bread-mould at the bottom of the scale, and the gigantic Wellingtonia of the Californian forests at the top! The one during the warm moist weather of summer appears suddenly, as if

by magic, on a stale crust laid aside in a dark
cupboard, attains its highest development, ripens
and scatters its seeds, and perishes in a few days ;
the other sent forth its embryo shoots in the
primeval solitude many ages ago, and may yet witness the revolution of many centuries ere it begins
to decay. The largest stalk of the Bread-mould
is no thicker than a pin, and may be half-a-line or
the twentieth part of an inch in height ; the trunk
of the Wellingtonia, like a huge church-tower, rises
nearly 300 feet into the sky, and measures upwards of a hundred feet in circumference. Why
does this enormous difference exist ? Why does
the fungus live for a day and the tree for ages ?
Why does one seed produce a plant that has but
a winter's or at most a summer's growth, and another grow into a plant which endures for more
than three thousand years ? They are both composed of the same materials—a collection and
combination of simple cells ; is it difference of
form only that gives a longer term of life to the
Wellingtonia than to the Bread-mould ? We cannot by any search ascertain the source of life in
the fresh seed, or account for the decay by which
mature development is followed, and there is
nothing in the structure of any plant, or indeed of
any created thing, out of which the assigned limit
of its life could be found. It is an impenetrable

mystery, to be referred humbly to the simple exercise of the Creator's will.

Fungi are extremely simple in their organization. They bring us back to first principles, and reveal to us the secret manner in which nature builds up her most complicated vegetable structures. They are composed entirely of cellular tissue, of a definite aggregation of loose, more or less oval, elliptical cells, with cavities between them. These cells in many species may be seen by the naked eye, and consist of little closed sacs of transparent colourless membrane. Here is the starting-point of life. Such cells are the primary germ or element from which every living thing, whether plant or animal, is produced. The whole process of vegetable growth is but a continuous multiplication of these cells. In the flowering plants the various vessels and organs arise in a *differentiation*, or a setting apart of particular groups of cells, and altering their forms and contents for the performance of particular functions in the economy of the plant. In the fungi, however, this specializing process advances but a little way. Their entire structure is uniform ; each group of cells is an exact repetition of all the other cells ; one part of each is exactly like the rest. There are no leaves, stems, or roots. Every cell is an assimilating surface ; the whole

plant is often only a reproductive organ. The common mushroom as gathered and brought to table
is merely the fruit of the fungus. It may be said,
however, of all fungi that they exhibit less or more
a well-defined separation into two parts, viz., the
mycelium or vegetative structure corresponding to
the thallus of the lichen which is horizontal, and
the fructification or reproductive structure which
is vertical; and just as in flowering plants the·
simple leaf, as is well known, is differentiated
into bract, calyx, corolla, etc., so among fungi
all parts are resolvable into modifications of the
mycelium.

Owing to this extreme simplicity and unity of
structure, they possess a remarkable power of reproducing and repairing such parts of their substance as have been injured. This power, it is
well known, is always more active as the organization of the individual, or the part affected, is less
complicated ; many of the simplest animals, such
as the polyps, admitting of being multiplied by
mere mechanical division almost to an unlimited
extent. It has been often remarked, that in man
and the vertebrata generally, the power of regeneration is confined to the replacement of small
portions of the simplest texture, although in them
the process of renewal is sometimes very extraordinary. The more highly organized structures,

such as muscular and nervous substance, cannot
be replaced ; should they be destroyed, the wound
is repaired by the formation of cellular, or some
other of the less complex tissues. Every part of
the fungus, however, as its structure is uniform
throughout, can be re-formed with equal facility.
Even the organs of reproduction, which may be
considered its most highly organized parts, can be
replaced or repaired if in any way injured. The
tubes of the toadstool, and the gills of the mushroom, have been cut out and separated from the
living plant by way of experiment, and yet in a
brief space of time they have been so carefully
reproduced, that no one could possibly tell they
had ever been removed. Snails are continually
eating holes into them, but when in active growth,
they speedily fill them up again with new tissue.
Puff-balls growing among grass on the borders
of woodlands, and in the open meadows, are
frequently very much injured by the scythe
of the mower, cut open, and whole parts sliced
off, but these wounds speedily heal themselves,
and the parts that have been removed are re-
modelled, without leaving the slightest cicatrice
to mark the point of junction or the seat of
injury.

Owing likewise to this extreme simplicity of
structure, they possess the faculty of almost inde-

finite expansion, determined only by the amount
of pabulum which the decaying substances on
which they are produced afford.   The limits of
some species are strictly marked out, and they
rarely exceed them, retaining nearly the same dimensions throughout their whole lives.   It is principally the smallest and simplest species which
are thus circumscribed ; and these make up by
their immense profusion for the insignificance of
their individual state.   The largest and most
highly developed species, which are but sparingly
produced, frequently attain to almost fabulous dimensions in favourable circumstances.   The scaly
polyporus (*Polyporus squamosus*), one of the commonest fungi, everywhere to be met with on the
decayed trunks of trees, especially the ash, and
easily recognised by its brown scaly pileus, and
white porous under-side, grows to a larger size
than any other species.   Instances have been recorded of its measuring seven feet five inches in
circumference, and weighing thirty-four pounds
avoirdupois, having attained these vast dimensions in the short space of three weeks.   The liver
fungus (*Fistulina hepatica*) has been found on an
ash-pollard weighing nearly thirty pounds.   Dr.
Badham, in his interesting work on the *Esculent
Fungi of Britain*, mentions having seen a fungus
in the neighbourhood of Tunbridge Wells which

rose nearly a foot from the ground, measured considerably more than two and a half feet across, and weighed from eighteen to twenty pounds. Some time ago a specimen of the *Polyporus annosus*, a species peculiar to pine-trees, was found in the Bank of England, Threadneedle Street, growing on a portion of the pitch-pine joists of the cellars. The entire growth was so large that when packed in a box for transit, it was as much as two strong men could carry. The largest piece was no less than seven inches thick and six feet three inches in circumference, and weighed thirty-two pounds ; while the piece of joist upon which it grew weighed only six and a half pounds. Specimens of agaric and puff-ball may frequently be met with, measuring a foot and a half in diameter, and weighing many pounds.

Although the structure of fungi is generally of a loosely cellular nature, yet they exhibit an astonishing variety of consistence. Each genus, and in many instances each species, displays a different texture. They range in substance from a watery pulp or a gelatinous scum to a fleshy, corky, leathery, or even ligneous mass. Some are mere thin fibres of airy cob-web, spreading like a flocculent veil over decaying matter ; while others resemble large irregular masses of hard tough wood. Their qualities are also exceedingly

various. Like the ferns they all possess a peculiar odour by which they may be easily recognised,
although it is somewhat different in different individuals, some smelling strongly of cinnamon
and bitter almonds, others of onions and tallow,
while others yield an insupportable stench. The
fœtid charnel-house smell of the common stinkhorn (*Phallus impudicus*) may be felt at a distance
of several hundred yards, when the wind is blow-
ing in one's direction, and leads infallibly to its
detection, when otherwise it might escape observation, covered, as it usually is, with leaves and
broken sticks. Like putrid meat it attracts flies,
which are always buzzing about its head ; and a
few individuals are sufficient to make a whole
wood intolerable. Bad as this species is, there is
another, if possible, in still worse odour—the
Clathrus, which is very rare in the southern parts
of our country, although abundant on the Continent. Like the curious leafless Stapelia, it dif-
fuses a most loathsome stench, which is utterly
insupportable at close quarters. This, with its
bright, coral-red network springing out of a white
gelatinous volva or egg, has originated the popular superstition among the peasants of the Landes,
that it is capable of producing cancer ; and hence
they cover it carefully over with leaves and moss
when they come across it in the pine-woods, lest

by accident some one should touch it, and be infected with the disease. As regards their tastes
the fungi are equally diversified, being insipid,
acrid, styptic, caustic, or rich and sweet. Some
have no taste in the mouth while masticated ; but
shortly after swallowing, there is a dry, choking,
burning sensation experienced at the back of the
throat, which lasts for a considerable time.

Endless variety of form, so constituted as to secure
a general uniformity of design and composition, is
the great characteristic of Divine workmanship ;
and nowhere is it more strikingly manifested than
among the lowest orders of plants. It is difficult
for us to conceive how simply, by a little change of
arrangement, and a little difference in the amount
and proportions of materials, such a countless
variety of objects can be produced,—objects,
though all composed of the same cellular tissue,
the same simple substances, yet so different in
appearance as to seem to have little or nothing in
common. And yet this is what is presented to us
in the great order of plants now under review.
Simple and uniform as is their structure, we have
seen how extensively diversified they are in
their specific qualities. They are no less varying in their forms. It is difficult to give a true
comprehensive idea of these varieties without
entering into specific details. Upwards of 3000

U

distinct species have been found and described in
Britain alone ; while more than twenty thousand
species altogether are known to the scientific
world. In round numbers it may be said that
fungi form about a third of the flowerless plants.
To show how numerous and varied are their
forms, it may be mentioned that the British species
are distributed in 368 genera,—an unusually large
proportion ; only eight species on an average being
included in each genus. A large number of these
species constitute separately distinct genera. In
no family of plants, indeed, are there so many
single forms, which, owing to the absence of affini-
tive characters, cannot be associated together,—
so many genera consisting of only one species.
While, on the other hand, there are no other plants
which have such immense genera, containing, some
of them, hundreds of species. The genus Agaricus, for instance, in this country alone has upwards of 450 species, so closely allied to the common mushroom of our tables, that many of them
are continually confounded with it, and yet exhibiting specific differences in colour, shape, size,
etc., so distinct as to be easily distinguished by an
educated eye. The two genera, Sphæria and
Peziza—whose ideal forms, in the former case a
simple round ball furnished at the apex with a
minute orifice, and filled internally with minute

flask-shaped seed-vessels ; and in the latter case, a shallow cup or plane disk of gelatinous matter, surrounded with a margin—are so diversified, that in Great Britain there are no less than 200 species of the one, and 166 species of the other. Some of the other genera are also unusually large, showing how rigidly nature's laws of uniformity and variety are adhered to in this class of plants.

The following instances may be brought forward, as illustrations of the remarkable shapes which many of the fungi exhibit. On the trunk of the oak, the ash, the beech, and the chestnut, may occasionally be seen a fungus, so remarkably like a piece of bullock's liver that it may be known from that circumstance alone. This is the *Fistulina hepatica* or liver fungus. Its substance is thick, fleshy, and juicy, of a dark Modena red, tinged with vermilion. It is marbled like beetroot, and consists of fibres springing from the base, from which a red pellucid juice like blood slowly exudes. Of all vegetable substances this exhibits the closest resemblance to animal tissue. Even in the minutest particular it seems to be a caricature of nature, a sportive imitation on an unfeeling oak-tree of the largest gland of the animal body. Tennyson might, with more truthfulness, personify an oak thus furnished with a substitute for the seat of passion, than the garrulous individual which adorned the woods of Sumner Chase! As already mentioned, it sometimes attains an enormous size, hanging down from the trunk of the oak like the liver of one of the geological monsters of the Preadamite world. Like the liver it is also nutritious, and forms a favourite article of food in Austria, though it is somewhat tough and acrid in taste. Another remarkable species of fungus, called Jew's Ears (*Hirneola Auricula-Judæ*) from its close resemblance to the human ear, clings to the trunks of living trees, particularly the elder, throughout the whole autumnal season. It is of a dusky or red-brown colour, like the ear of a North American Indian, and is wrinkled with large swelling veins branching from the middle, where they are strongest, and somewhat convoluted, the upper side covered with a hoary velvet down, the inside smooth and darker coloured. When it grows on a perpendicular stump or tree, it turns upwards. Another remarkable species, the *Tremella mesenterica* (Fig. 27), common all the year round, on furze and sticks in woods, bears a strong resemblance to the human mesentery. It is of a rich orange colour. This extraordinary resemblance which different fungi bear to the different parts of the animal body, served to confirm the opinion of the ancient botanists and herbalists, that they were animal structures, or at least intermediate links between the animal and vegetable kingdoms.

The simplest fungi consist of a few primordial cells, either separate or conjoined, or of cellular, branched filaments or threads, performing the functions of nutrition and reproduction. Between these and the mushroom, which may be regarded as exhibiting the highest development of fungoid life, there are numerous intermediate forms more or less complex. Some resemble minute mussels

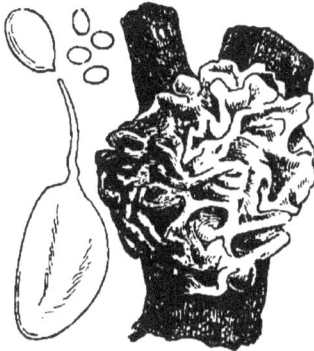

FIG. 27.—TREMELLA MESENTERICA.

with their edges upwards ; some are shell-shaped, and others shrubby and branched like coral. Some form large round balls, splitting into star-like expanding rays ; others are crowned with mitres or peaked caps. Some are cup-shaped, trumpet-shaped, bell-shaped. Some, such as the leaden-coloured *Crucibulum vulgare* (Fig. 28), so frequent on rotten wood and in potato-fields, form a nest in which to rear their young. One forms a yellow scum on moss-tufts in woods, which in a few days dries up and becomes converted into a heap of black powder like soot ; another forms, on the stems of grass some inches above the soil, a thick white froth, somewhat resembling the salivaceous exudation of the *Cicada spumaria* so frequent in summer woods, and which may easily be supposed of animal origin.   Some form beautiful little goblets elevated on slender hair-like stems ; while others are only to be seen through a thick red lattice-work which surrounds them.

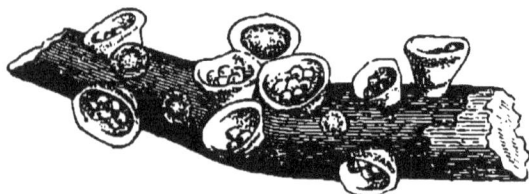

FIG. 28.—CRUCIBULUM VULGARE.

In short, there is almost no end to the curious shapes which this tribe exhibits. Nature, in a capricious or sportive mood, seems to have formed them in imitation of the higher objects of creation, as they are her humblest and latest productions. Having such extremely simple and plastic materials to work upon, she seems to have followed the wildest vagaries of fancy in the determination of their shapes, and to have moulded many of them in imitation of the substances upon which they are produced.

Although fungi in general are sober, nun-like

plants, preferring quiet quaker colours suitable to
the dim secluded places which they usually affect,
yet some of them depart widely from this sober-
ness, and exhibit themselves in the most gaudy
hues. Some species are of a brilliant scarlet
colour; others of a bright orange. Many are
yellow, while a few don the imperial purple. In
short, they are to be found of every colour, from
the purest white to the dingiest black, dark
emerald or leaf-green alone excepted. Some are
beautifully zoned with · iridescent convoluted
circles, or broad stripes of different hues. Some
shine as if sprinkled with mica; others are smooth
as velvet, and soft as kid-leather. Such is a rapid
survey of the varied forms, colours, and qualities
exhibited by these simple plants; and surely it is
sufficient to show us the vast amount of interest
connected with them.

Let us take a specimen of one of the most
perfectly-formed and highly-developed fungi, the
common shaggy mushroom for instance (*Agaricus
procerus*, Fig. 29), which is also the most familiar
example, and endeavour to point out the peculiarities of its structure. Like all plants, it consists
of two distinct parts, the organs of nutrition or
vegetation, and the organs of reproduction; the
former bearing but a very small proportion in
size to the latter. The organs of nutrition or

vegetation consist of greyish-white interlacing filaments, forming a flocculent net-like tissue, and penetrating and ramifying through the decaying substances on which the mushroom grows. These filaments are formed of elongated colourless cells. They are developed under ground, and in other plants would be called roots. This part of the fungus is called by botanists mycelium, and is

FIG. 29.—PARTS OF MUSHROOM (*Agaricus procerus*).
(*a*) Pileus or Cap. (*b*) Hymenium or Gills. (*c*) Annulus. (*d*) Stipe or Stalk. (*e*) Volva. (*f*) Mycelium or Spawn. (*g*) Spores. (*h*) Basidia.

popularly known as the spawn by which the mushroom is frequently propagated. In favourable circumstances this mycelium spreads with great rapidity, sometimes, especially when prevented from developing organs of reproduction, attaining enormous dimensions. It may be kept dormant, in a dry state, for a long time, ready to grow up into perfect plants when the necessary

heat and moisture are applied. When the requisite conditions are present, and the mycelium
begins to develop the reproductive tissue, there is
formed at first a small round tubercle, in which
the rudiments or miniature organs of the future
plant may after a while be distinctly traced. In
this infantile condition, the mushroom is covered
completely with a fine silky veil or volva, which
afterwards disappears. The tubercle rapidly increases, until at last it produces from its interior
a long, thick fleshy stem or stipe, surmounted by
a pileus, or round convex, concave, or flat cap,
similar to that anciently worn by the Scottish
peasantry. This is the organ of reproduction,
equivalent to the thecæ of mosses and the flowers
of phanerogamous plants. This cap is covered
with a veil or wrapper, which is ruptured at a
certain stage, and retires to form an annulus or
ring round the stem. When it is removed from
the under side of the pileus, a number of vertical
plates or gills is revealed of a pale pinkish-yellow
or white colour, different from the rest of the plant,
and radiating round the cap from a common centre.
The whole of this apparatus is called the hymenium.
Each of the gills, when examined under the microscope, is found to consist of a number of elongated
cells called basidia, united together on both sides of
a cellular stratum, and bearing at their summits

314 FIRST FORMS OF VEGETATION.

four minute spores supported on tiny stalks. It
is by these spores which become detached when
ripe that the plant is propagated. When a small
fragment of a ripe gill is placed on the glass slide
of the microscope, in a drop of water, the spores
will detach themselves from the gill and float
freely on the water ; or even if a whole mushroom
be laid on a sheet of paper, it will often leave
behind its spores in the form of a thin impalpable
powder. These spores are so very minute, that
many thousands of them are required to make a
body the size of a pin-head ; and they are capable
of enduring a temperature at least equal to that of
boiling water, as was satisfactorily proved a few
years ago when the barrack bread in Paris was
affected with mould, which was in active growth
almost before the bread was cold. They are also
of different colours, being white, rose-coloured,
brown, purple, and black; and this variety of
spore-tints affords a ready test for grouping
species of Agarics, which by a little practice will
become easy. The spores and sporidia of fungi
have a singular tendency to appear in definite
numbers, either in twos, fours, or multiples of
four. Every observer is struck indeed with the
quaternary arrangement which prevails in all
cryptogamic plants, while few facts are more
curious than that the number four should prevail

when the fructification consists of naked spores, and a multiple of four when it is contained in tubular cases or asci. Along with the basidia or formative cells of the hymenium occur large sterile cells, flask-shaped, and containing granular matter exhibiting molecular motions when discharged. These are called *cystidia*, and were supposed to represent antheridia, but they are now ascertained to be mere abortive basidia. Just as the petal or carpel of a flowering plant changes abnormally into a green leaf, so the cystidia of the mushroom abandon the reproductive and return to vegetative functions by a sort of hypertrophy.

While upon the subject of spores I may mention here that the remarkable elastic force with which many of the fungi eject their seed has often excited attention, and is fully equal to anything of the same kind observed among flowering plants. In hot-houses, adhering to decaying leaves, may occasionally be seen a curious little plant called *Sphærobolus stellatus* (Fig. 30), which bears no inapt resemblance in its shape and functions to a Liliputian mortar. It is of a pale straw-colour, and consists of two coats, both stellated, and separated from each other by a bead of dew exuded by the plant. The rays of the outer case are orange. No sooner is the inner case touched, than it becomes suddenly inverted, and shoots forth, with a jerk, a little pellucid ball to a distance of upwards of three feet. This ball or sporangium contains the seeds, and is ejected with a force which, considering the nature and diminutive size of the plant, far exceeds that employed in the projection of a shell from the largest mortar, or a cannon-ball from an Armstrong gun. It is a far more curious and interest-

FIG. 30.—SPHÆROBOLUS STELLATUS.
Natural size, and magnified.

ing object than the squirting cucumber. Another denizen of the hot-bed (*Peziza vesiculosa*) exhibits somewhat similar properties. When the sun is shining warmly upon its cup, the least agitation raises a visible cloud of sporidia like a thin wreath of vapour. These are beautiful instances of the adaptations, with which nature has provided these lowly plants, for the certain dissemination of their seed

The mushroom may be regarded as an ideal fungus of the highest type ; and consequently the preceding description is only applicable to the class which it represents. There are varieties of structure as there are varieties of form. There are six large orders of fungi in which the organs of fructification are widely different. The first order is called *Hymenomycetes*, or naked fungi, because the seed-bearing organs are naked, or placed externally. This is the largest, most important, and most highly developed order. The mushroom, toadstool, chantarelle, amadou, are familiar examples of it. The hymenium assumes various shapes in the different genera. In the mushroom it forms gills, in the toad-stool tubes, in the chantarelle veins, in the amadou pores, and in the hydnum spines. The second order, called *Gasteromycetes*, has the seed-bearing organs enclosed in a membranous cover-ing, like the stomach of an animal, whence the name. The stinkhorn, the Melanogaster or red truffle of Bath, the bird's-nest fungus, and the puff-ball, are familiar examples of this order. Some of the forms, such as *Stemonitis fusca* (Fig. 31), common on rotten wood, are ex-ceedingly elegant. The third order is called *Conio-mycetes* or dust-fungi, because the spore-cases are produced beneath the epidermis of plants, or the

matrix in which they are developed, in the form
of a minute collection of dust, entirely destitute of
any covering or receptacle, except that which is
furnished by the skin of the plant raised around
them. This class is the most
destructive of the whole tribe.
Smut, bunt, and rust, are *too*
familiar examples of this most
notorious class. The fourth
order is called *Hyphomycetes*
or web-like fungi, because the
spores are free, developed on
naked filaments whose terminal cells are often transformed into a series of spores
like a row of beads. The
general appearance of the
plants belonging to this order
is that of a quantity of dust-
like seeds imbedded in a
flaky cottony substance, like
a spider's web. The different
kinds of common mould,
blue, yellow, and green, the
potato-disease, caterpillar and silkworm blights,
and various kinds of mildew, are common examples of this order. The limits of the genera and
species belonging to the orders Coniomycetes and

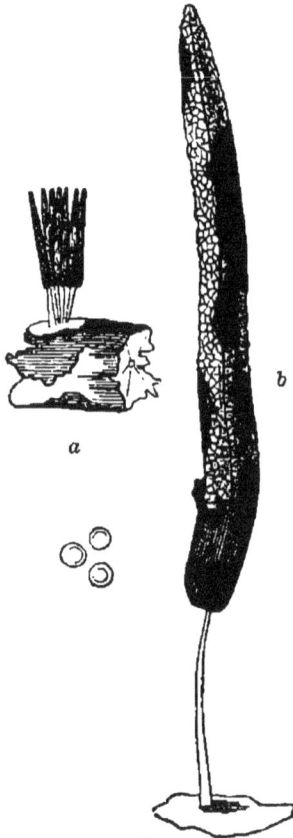

FIG. 31.—STEMONITIS FUSCA.
(*a*) Group of natural size.
(*b*) A single individual magnified.

Hyphomycetes are however somewhat indefinite. It is indeed doubtful how many of them are auto- nomous. Numerous forms are truly polymorphic, appearing under different phases ; so that what were formerly regarded as good and distinct species are now found to be in reality only con- ditions or immature stages of other forms. The fifth order, called *Physomycetes*, is distinguished by its stalked sacs containing numerous spores or sporidia. It is the smallest of all the orders num- bering only a dozen British genera, and about two dozen species. The black felty cellar-fungus, and the grey mucor or mould on preserves are familiar illustrations of this order. The sixth and last order is that of the *Ascomycetes* or asci-bearing fungi, whose spores, generally eight in number, are pro- duced in the interior of groups of elongated sacs or thecæ contained in fleshy, leathery, or wart-like fructification. These fungi, of which the morel, truffle, and vine disease are well-known examples, numbering in this country about a thousand dif- ferent kinds, resemble lichens in every respect ex- cept that they are produced on decaying sub- stances, and are possessed of a mycelium or spawn destitute of the green cellular matter of lichens. By some authors, such as Schleiden, they are in- cluded among the lichens notwithstanding these discrepancies. The accompanying illustration of

the *Marsh Mitrula* (Fig. 32), of a bright orange-
yellow colour growing abundantly in July on dead
leaves in wayside pools, will give a good idea of
this order. Such is a brief analysis of the different orders of British Fungi, and a general survey
of the different kinds of fructification.

We find the same gradation in the scale of development among the fungi that exists among the

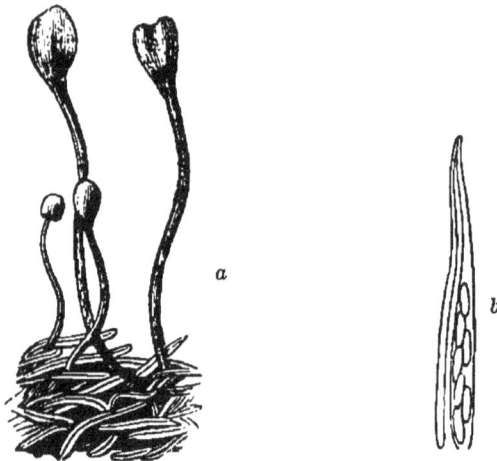

FIG. 32.—MITRULA PALUDOSA.
(*a*) Natural size. (*b*) Ascus containing sporidia, highly magnified.

flowering plants. The moulds and mildews are
analogous to annual herbaceous plants, and the
agarics represent trees. And just as a tree is not
an individual but a perennial colony of annual
plants, growing vertically in the air instead of
horizontally on the ground, so a mushroom may
be theoretically regarded as a mass of closely compacted mould, a corporate structure built up of a

number of individual fungi that grow vertically together instead of horizontally in a crust or tuft, and thus endowed with greater powers of endurance and longevity. In the mushroom it might be possible to separate a filament from the spawn, and trace its course up the stem, through the cap, down one of the gills to the surface, where it bears sacs or basidia with their four naked spores. And this isolated fruit-bearing thread of the mushroom would be equivalent to an individual mould-plant, with its free filaments bearing terminal or lateral spores. It is therefore in strict accordance with morphological rules to consider the whole mushroom as typically an aggregation of separate individual fungi of lower type, consolidated together into one individual. And as if to favour this view, we find that, under certain circumstances, the common green mould, *Penicillium crustaceum* (Fig. 37), instead of forming, as it usually does, a continuous stratum of separate individuals, breaks up into little tufts, the threads composing which are so incorporated as to form a sort of common stem with a globose head of spores, thus making an approximation to the stem and cap of the mushroom. It is thus exceedingly interesting to find the same laws of morphology that control the highly organized forms and structures of the trees of the forest, applicable to the humblest mould

X

that creeps over an old shoe and that perishes in
a day.

Reproduction among the fungi is not so simple
a process as was formerly supposed. Sexual elements have been found among them, as among all
other cryptogamic plants. Spermogones (Fig. 33)
or antheridia, entirely similar or analogous to those
of lichens, are observed in different tribes. In the
cavity of these spermogones, filled with a viscid

FIG. 33.—SPERMOGONES OF ÆCIDIUM BERBERIDIS.—Highly magnified.

fluid, occur very minute cylindrical bodies called
spermatia, similar to the so-called spermatozoids of
lichens. These exhibit the characteristic movements already described, evolve a perceptible
odour, which has been compared to that of the
pollen of the willow, and when ripe are expelled
from a small orifice in the summit of the spermo-
gone. Zoospores swimming off when disengaged
from the conidia, by means of vibratile cilia, have

also been discovered in some of the white rusts and parasitic moulds. In bunt it has been ascertained that an alternation of generations takes place ; while in the higher fungi curious instances of conjugation, like that which occurs in fresh-water algæ, are on record. The mode of growing agarics from the spores, however, is still involved in obscurity. No one can say whether the stem and cap rise at once from the mycelium without any sexual process, or whether the spore of the mushroom produces a prothallus similar to that of ferns, on which sperm and germ cells are developed, giving birth by their union to the full-formed plant. We cannot by any known method grow domesticated mushrooms from spores ; and in a wild state the myriads of spores that fall to the ground where they grow, may yield next season not a single specimen. They behave like unimpregnated germs or ova. The conditions that are needed to make them fertile are unknown. Indeed, the physiological relations of the various seemingly reproductive structures which I have described are as yet quite obscure, and present a field for most interesting observation and research. But it is worthy of remark that in the lowest cryptogams the mode of reproduction far more closely resembles that of the highest animals than what we see in flowering plants. And the curious

thing is, that instead of locomotion being the distinguishing mark of the most advanced stage of
growth in these organisms, it is peculiar, as we
have seen, to the infantile condition ; while the
adult approaches more to the Buddhist idea of
perfection, and settles down to repose.

From the preceding observations it is evident that
all the forms of fungoid life, excessively minute in
size and simple in structure although many of them
are, obey the great law of nature in propagating
themselves by seeds or germs. And yet there are
not wanting individuals who believe that these
plants are the productions of spontaneous or equi-
vocal generation, springing up without seed or germ
from the soil, or from substances in a state of fermentation. This theory is countenanced and rendered
plausible by the almost instantaneous appearance
of mildew, dry-rot, mould, and various others of the
simplest class of fungi on the objects affected, and
the strange and almost inaccessible situations in
which they are found, as, for instance, in the inside
of a large cheese, in a hazel-nut, in a fresh egg, in the
core of an apple, beneath the wrapper with which
the careful housewife covers her cherished preserves, and under the epidermis of living plants,
—localities where it is difficult to conceive how
any seed, however minute, could find lodgment.
The nature and habits of these plants are now,

however, better understood than they formerly
were ; and although the controversy is still
going on with unabated interest between the ad-
vocates and the opponents of the doctrine of spontaneous generation, the balance of proof appears
to me to be decidedly on the side of the latter.
The beautiful researches of M. Pasteur, models
alike of scientific experimentation and logical
reasoning, have, as far as I can judge, established
the fact, that a seed is as necessary for the production of the minutest speck of mouldiness which
the microscope can reveal to our view, as the acorn
is for the germination of the giant oak of the
forest, or the date for the growth of the magnificent palm of the desert. It is true that fungi are
most frequently found on the products of animal
or vegetable decomposition ; but they occur in
such situations, not because these decaying substances originate them, but because they afford
them the necessary conditions of their growth,
their germs having been previously deposited there
by pre-existing species. If we sow a quantity of
the spores of the common bread-mould on a stale
crust, we shall have a quicker growth and a more
abundant crop of fungi than if the crust be left to
a natural or chance supply of seeds ; just as the
farmer has a surer and more plentiful harvest
when he deposits a sufficient quantity of seeds in

the ground, than when he leaves the chance of a crop to the scattering self-sown wheat of the previous autumn. Indeed, the closest and most prolonged observations, and the most carefully-conducted experiments—and some of them, especially those of Dr. Bastian, have been conducted with consummate skill and patience,—have not led to the proof of a single instance of spontaneous generation, even of one of the simplest of all living things ; but, on the contrary, they all, in my judgment, lead farther and farther from, or entirely disprove it. I believe, from the results already obtained, that if due care be taken to get quit of the ova of animals, and the seeds of minute vegetables from any fluid or other suitable matrix, and at the same time carefully to exclude the further entrance of them through the admitted air, no traces of animal or vegetable life will appear. The presence of mould in such an apparently inexplicable place as the interior of a large cheese, is owing to the exposure of the curd to the air when the cheese was being made, and the consequent deposition upon it of the minute germs of fungi floating around, which afterwards developed themselves when the curd thus impregnated formed the inside of the cheese. It is well known that the exposure of curd for a single day to the atmosphere will have the effect of producing mouldy cheese. The same may be said regarding the existence of fungi in the inside of an egg, hazel-nut, or apple. Professor Panceri inoculated a fresh ostrich egg at Cairo with a fungus, thus proving that its shell is permeable by spores.

Countless millions of the subtle seeds of fungi, invisible to the naked eye, and light almost as the particles of vapour around them, are continually floating in the air we breathe, or swimming in the water we drink, or lying amid the impalpable dust and sand of the soil, waiting but the combination of a few simple circumstances, the presence of warmth or moisture, or a suitable matrix, to display their vital energies, and to burst into full, free, independent life. Myriads of the minute germs of the various moulds which approach us in our houses, and fasten upon different articles of domestic use, may be and often are dancing about in the air-currents of our apartments, though totally invisible to us ; but could we sufficiently magnify them, as a sunbeam darted in at our windows and illuminated their bodies, they would appear like so many cannon-balls, moving rapidly up and down, and in every direction. The microscopist and the chemist have demonstrated the existence of these germs in greater or less quantity in the air of the country as well as in the air of the town, out-of-doors as well as in-doors ; and

Professor Tyndal, by calling in the aid of optical analysis, has made assurance on that point doubly sure. If we venture for a moment to imagine the overwhelming number of seeds which the different species of fungi must disseminate in the course of a single year,—if we consider that each individual of the common puff-ball contains upwards of ten millions of seeds, and these so small as to form a mere cloud when puffed into the air, and that a single filament of the mould which infests our bread and preserves will produce as many germs as an oak will acorns, so that a piece of decaying matter, not two inches each way, will scatter upon the air, at the slightest breath of the summer breeze, or the gentlest touch of the smallest insect's wing, as many seeds, quick with life, as this country will produce of acorns in a twelvemonth ;— if we take these things into consideration, it is not too much to suppose that the seeds of fungi must be ubiquitous, and from their excessively minute size penetrate into every place, even into the stomachs and other parts of animals. Indeed, the difficulty seems to be rather to imagine a spot altogether destitute of them than to account for their universal diffusion. This circumstance has been made the ground of a belief that malarial and epidemic fevers have their origin in cryptogamic vegetables or spores. Much valuable information has of late years been acquired regarding the habits and modes of propagation of these diseases ; but little as yet has been ascertained regarding their essential nature. The pestilence still ' walks in darkness,' and neither chemistry nor any other science can tell us what is its essential nature, nor in what its terrible potency consists. If the spores of fungi be really the exciting cause, in predisposing circumstances, of zymotic diseases, these minute bodies, conveyed through the air, and introduced into the body in respiration, could easily be detected. Professor Fries has compared the relative magnitude of a large proportion of fungoid sporules to that of the globules of chyle and blood in the human subject, although many are about two-thirds of the size of the former and one-third that of the latter ; while particles of inorganic matter can be distinguished by the microscope so minute as the 200,000th part of an inch. Be the origin of these diseases, however, what it may, it is a matter of fact that when cholera appeared in this country, in 1847, an extraordinary quantity of these microscopic spores were found in the air. If they were poisonous, as many of the fungi are, or were capable, in the manner of a ferment, of exciting morbid actions in the system, it admits of being suggested at least that those living in places where dense clouds of them were present, being devitalized by other noxious influences, such as vitiated air, defective sewerage, bad water, or an inadequate supply of food, and consequently in a state of body unable to resist the deleterious action of these cryptogamic germs, died from a form of poisoning. These countless myriads, then, of invisible seeds which continually float in our atmosphere, ever ready to alight and spring into life, as the advanced heralds of the plague and the pestilence, may well strike us with astonishment, if not with awe. Above us, about us, and in us, they roam like vigilant spirits, seeing that all is right with our physical constitution ; but availing themselves of the slightest flaw to work our destruction.

Although fungi are in an especial manner capable of universal dissemination, yet we find that in their geographical distribution they are as much restricted as other plants. Some representatives of the class are found in every part of the world, and some particular species have the power of indefinite extension and localization, but, as a whole, like the higher cryptogams, they can only spread within certain limited areas. The habits of fungi convince us that there is something apparently capricious in their distribution, or rather, that some only of the conditions upon which their multiplication depends are at present known. In

tropical forests, where the exuberance of the vegetation excludes the rays of the sun, and creates the
dim light and the still, moist air which they love,
and where there is always an immense quantity
of decaying organic matter, we might expect to
find them in the greatest quantity and luxuriance.
But, strange to say, fungi as a class are comparatively rare in tropical woods.   While every tree
has its creeper, and almost every flower its parasite, the plants which, above all others, are most
parasitical have relatively few representatives
there ; and dead trunks and prostrate boughs and
decaying herbage rot and crumble away untouched by the ravages of mushroom or mould.
Insects in these countries perform the office of
fungi in hastening the decomposition of dead
matter, and incorporating and deodorizing the decaying particles ; and it must be confessed that
they perform this duty more speedily and effectually ; while, unlike the fungi, they leave no unpleasant traces, no putrifying masses behind when
their work is accomplished, and their own turn
comes to die.   Like some of the epidemic diseases,
as, for instance, typhus, with which they are said
to be connected, the too high temperature of the
tropics seems to offer an effectual barrier to
their general distribution in those countries.
Their head-quarters seem to be in northern lati-

tudes, where the temperature is mild and genial,
and where there is a constant supply of moisture.
Professor Fries of Upsal, the presiding genius of
these plants, gathered in Sweden, within a space
of ground not exceeding a square furlong, more
than two thousand distinct species. ' This country,' says Mr. Berkeley, 'with its various soils,
large mixed forests, and warm summer temperature, seems to produce more species than any part
of the known world ; and next in order, perhaps,
are the United States, as far as South Carolina,
where they absolutely swarm. A moist autumn
after a genial summer is most conducive to their
growth, but cold, wet summers are seldom productive. The portion of the Himalayas which lies
immediately north of Calcutta is perhaps almost
as prolific in point of individuals as the countries
named above, but the number of species on examination proves far less than might at first have
been suspected. It is probably not a fifth of what
occurs in Sweden. Great Britain, though possessing a considerable list of species, is not abundant
in individuals, except as regards a limited number
of species. The exuberance, even in the most
favourable autumn, is not to be compared with that
of Sweden or many parts of Germany.' They are
found in the Arctic and Antarctic regions, almost
as far as the limits of vegetation. They penetrate

to the dreary regions of Greenland and Lapland,
supplying the natives with their tinder, and with
an excellent styptic for stopping blood and allay-
ing pain ; and they announce to the hapless exiles
of Siberia, when their gaily-coloured forms spring
forth from the crevices of the rocks, and in the
dark haunts of the gloomy fir-woods, that the
stormy blasts of winter and spring are past, and
that the summer and autumn, those short, sweet
seasons of indescribable beauty and pleasure, have
come.

Certain genera and species occur only in tropical
and sub-tropical regions, having their northern
limit in the north of Africa or the coast of the
Mediterranean. Several genera and species are
confined to New Zealand, others to Ceylon and
Java, others to the Cape de Verde Islands and the
United States. Like flowering plants, the fungi
of different climates and zones are found at different heights along the sides of tropical mountains
that rise above the snow-line. In the Sikkim
Himalayas, *Polyporus sanguineus* and *xanthopus*
luxuriate in the stifling tropical woods at the
base of the hills ; higher up the fungi peculiar
to Ceylon and Java grow among the palms and
tree-ferns of the mid regions ; higher still, the
species of southern Europe abound in the deodar
forests and among the rhododendron thickets of

the upper heights; while below the line of perpetual snow, on grassy slopes and amid shrubby vegetation, may be seen species, if not identical with, at least very closely allied to, those of Britain and Sweden. One species has been found at a height of 18,000 feet, which is probably the highest range of fungoid life.

But while the fungi are, to a certain extent, restricted in their geographical distribution within certain well-known limits, they are, on the other hand, almost ubiquitous in their choice of habitats. There is hardly a single substance on which some species or other of them may not, under favourable circumstances, be found. ⌃ As a general rule they all grow on dead and decaying organic matter, on the mouldering trunks and branches of trees and withered plants, and on the bones and droppings of animals. But they are also occasionally found on living trees, and on green leaves, and parts of plants that show no symptoms of decay. A large class called hypodermous or entophytic fungi spring from beneath the cuticle of living plants. There is hardly a single flowering plant which is not infested by them—a different fungus being developed upon almost every species. Their minute sporules are either directly applied to the plants upon which they are found, entering by the stomata or breathing pores; or they are

taken up from the soil by their seeds in the process of germinating, enter into their structure, cir-
culate through their tissues, remaining all the time
in a dormant state, until at last, when the part
which forms the most suitable nidus for them is
developed, they suddenly appear upon it externally in the form of patches or aggregations of
black or coloured granules. A very interesting
example of these epiphyllous fungi may be seen
on the wood-anemone. It is called *Puccinia ane-
mones*, and is one of the earliest and commonest
species. No sooner does the true foliage, proceeding from the rhizome of the anemone, appear above
the sod in early spring than several of the leaves
are seen to be sickly-looking, attenuated, pre-
maturely developed and rising higher than the
others. On examination their under surface is seen
to be covered with brown spots, which bear so
close a resemblance to the sori of ferns that the
infected plant used to be classed among ferns, and
is still considered to be such by neophytes. In
Ray's time it was known to botanists as the Con-
jurer of Chalgrave's Fern, and ranked with the
maidenhair and polypody as a rustic remedy.
Another example of fungi growing on living
plants is familiar to the farmer under the name of
'berberry blight,' *Æcidium berberidis* (Fig. 34).
The theory has long been prevalent among prac-

tical agriculturists that wheat in the neighbourhood of a berberry bush seldom escapes the
blight.   Long regarded by scientific men as a

FIG. 34.—ÆCIDIUM BERBERIDIS.
(*a*) Branch of berberry with spots of rust. (*b*) Natural size. (*c*) A portion of the rust, highly magnified. (*d*) Sporidia, highly magnified.

prejudice of the bucolic mind, this theory has of
late received remarkable confirmation by the dis-

covery of dimorphism, or the phenomenon of the alternation of generations among fungi. The well-known orange-red spots so common on the leaves of the berberry are but one form of the yellow rust of wheat and other cereal crops. If the leaves of the berberry are inoculated with the spores of the wheat-rust, they will produce the Æcidium or berberry-blight, figured above ; and if the spores of the Æcidium are sown on the leaves of wheat, they will produce the wheat-rust. But, strange to say, if the spores of either form be sown on the plant on which it is itself parasitic, they fail altogether to produce the same plant ; and this alternation of generations will account for the fact, so long considered mysterious, that rust is apt to appear not in successive but in alternate years on the same crop. The class of fungi to which the berberry-blight belongs is exceedingly beautiful, and abounds in most interesting objects for low powers of the microscope.

Many fungi, contrary to the habits of the race, seem to live on mineral matter. Numerous exotic Polypori, for instance, grow on hard volcanic tufa, without a particle of organic matter. Other fungi are not unfrequently found in this country growing in abundance on the hardest gravel stones, and bare plastered walls destitute of all organic nourishment. Mr. Ivor found a Didymium on a

Y

leaden cistern at Kew ; another was found by Mr. Sowerby, in the outer gallery of St. Paul's, on cinders ; while a still more extraordinary instance is related by Schweinitz of a species of Æthalium vegetating on iron which had been subjected to a red heat a short time before. 'A blacksmith,' he says, 'at Salem, by no means void of sense or cultivation, had thrown on one side a piece of iron which he had just taken from the fire, being called off to some other business. On his return in the morning, he was astonished to see on this very piece, lying over the water in his smith's trough, a quantity of this fungus, of a soft gelatinous consistency. He immediately sent for Schweinitz without moving anything from its place, who was equally astonished to find a distinct species of Æthalium. The mass of fungi was two feet in length, consisting of a series of many confluent individuals. It had crept from the iron to some adjacent wood ; and not, as might be objected, from the wood to the iron. The immense mass had grown in the space of twelve hours.' This plant forms a yellow pulpy mass, like curdled egg, in tan-pits and hot-houses, cucumber and melon frames, where it is very common and injurious. It is also found on cinders and lead. In the woods it grows on mosses. All these curious instances show that fungi do not always derive nutriment

from their matrix, and that some of them are es-
sentially meteoric, depending on matter conveyed
to them by the surrounding air or moisture. A
species of Phycomyces, which bears a strong resemblance to an alga, from its green colour and
shining aspect when dry, grows rapidly and in prodigious quantities in soap and candle manufac-
tories, covering walls that are saturated with oil
or grease in immense flakes. In tallow casks it
flourishes in the most wonderful degree, and ultimately exhausts the grease on which it grows to
a great depth. It is supposed to be a transformation of the common green mould. Some
species, such as the truffle, are subterranean, vegetating in the absence of all the external stimulants
upon which other plants depend, being apparently
attached to the roots of trees, often at a considerable distance underground. Some species are
found clinging to the roof of coal mines, and hanging from it in fantastic fashion in masses, snowy
white as a sheep's fleece, but turning to a disagreeable brown paste as soon as handled. A peculiar
fungus (*Zasmidium cellare*), like a bacchanalian
gnome, is found on casks and bottles, and hanging
down from the roof in close cellars. It grows in
great abundance in the London docks. The dim
vaults, with their vistas of casks, extending in the
darkness farther than eye can reach, are festooned

with this fungoid cobweb, hanging from the roof
like a soft and comfortable form of stalactite, in
the strangest forms and in immense masses. It
begins as an incrustation resembling white cotton
wool on the brickwork of the vault, and as it
grows, descending in irregular shapes, hanging
down a foot or two in length, and changing to
a dingy brown colour, very like a mouse-skin.
The men who live in the place are proud of this
extraordinary fungus, which carries out the convivial aspect imparted by the wine casks; it is
never interfered with, and they point out any
larger mass than usual with some complacency.
Many fungi prey upon other fungi. The parasitism of the mistletoe among flowering plants is
paralleled among flowerless by the *Nyctalis asterophora* (Fig. 35), which, itself an agaric, grows on
other agarics.

‹ As a singular instance of the ease with which
these plants can accommodate themselves to surrounding circumstances, it may be mentioned that
several species of fungi of the genus Chionyphe,
somewhat allied to the common moulds of our
cupboards, are found growing upon melting snow.
They are among fungi what the red-snow is
among algæ. One of these *snow moulds* was
first discovered in the north of Iceland ; but two
other species have since occurred in Germany in

great abundance. The Chionyphe is developed on the snow in clear weather, when the sun has power enough to melt the upper crust, without the existence of a general thaw ; and in all probability springs from the droppings or the urine of animals decomposed in the snow. It spreads over the surface of the snow in shining fleecy patches, dotted with red or green particles. When the snow melts, it is left behind upon the underlying

FIG. 35.—NYCTALIS ASTEROPHORA GROWING PARASITICALLY ON DEAD RUSSULA NIGRICANS.
(1) Section, natural size. (2) Stellate spore, highly magnified.

grass in the form of a cobweb stratum, which in a few days disappears. Another species of snow-mould discovered in Germany, and described by Professor Unger, under the significant name of *Lanosa nivalis*, unlike the former, grows underneath the snow ; and in certain seasons, especially when a deep snow sets in without any previous frost, it is extremely destructive to the grass upon

which it is developed. It forms white patches a
foot or more in diameter, made up of a number of
smaller circular patches, which assume here and
there a red tint, as if dusted with red powder.
When the snow melts at the approach of spring
the whole disappears, leaving behind a withered
plot, which, according to the greater or less vigour
or duration of the parasite, is either completely
barren, or but slowly resumes its verdure.

Not content with preying upon dead organic
matter, or growing plants, some fungi also attack
living animals. In this country there are several
species of Torrubia, particularly the *T. militaris*,
distinguished by its fleshy orange-red club-shaped
appearance, which grows in garden soil on the
larvæ and pupæ of insects ; while others are
parasitic on the *Elaphomyces granulatus* and
muricatus in pine woods. In New Zealand, a
remarkable species long known as the *Sphæria
Robertsii* grows from the head of the caterpillar
of the *Hepialus virescens*, a kind of moth, when it
buries itself among the moss in the woods to
undergo its metamorphosis. In appearance it is
a somewhat crooked, long, slender stalk, terminating in the spike-like fructification. Its growth
destroys the caterpillar ; the grub, instead of
developing itself into a beautiful butterfly, being
replaced by a nauseous fungus. It is so common

and prominent a species in New Zealand, that it
has a name in the native language, and is associated with some of the ancient Maori superstitions.
In the West Indies, wasps called by the inhabitants *Guêpes végétantes*, may often be seen
flying about with fungoid plants as long and
nearly as bulky as their own bodies growing upon
them ; while in this country itself, it is by no
means rare to see a humble-bee, or a common
blue-bottle fly, that had been killed by the growth
of a club-shaped Sphæria from its body, from
half an inch to an inch in length, of a sienna
brown or lemon colour. In the forests of Pomer-
ania and Posen the caterpillars have been destroyed in enormous quantities during certain
seasons by a fungoid epidemic, caused by the
mycelium of *Empusa Aulicæ*. It attacks every
order of insects, with the exception of dragon-flies,
in all stages of growth, and develops with prodigious rapidity in the individual. Flies are
usually attacked by a fungoid disease about the
end of autumn, when the cold damp weather
which then prevails has reduced the vitality of
their bodies to the lowest point, and rendered
them incapable of resisting external agencies.
At this time they forsake their accustomed haunts
in the open air, congregating within doors for
warmth and shelter, and may be seen in consider-

able numbers adhering to window-sills, walls, and
various articles of furniture in our rooms. In a
few days they die, but strange to say their appearance is so little altered, that it is impossible without actual examination to tell that they are dead.
When dying in the ordinary way, they always
draw up their legs, and cross them beneath their
bodies ; but when they perish of this disease, the
legs are stretched out supporting their bodies, and
retaining them in their natural position. The pro-
boscis is protruded, as if in the act of imbibing
nourishment, and their whole appearance is that
of vigorous healthy flies that have alighted for a
moment, and may be expected in the next to take
wing and fly away. The only difference observed
is a whitish halo, like a sprinkling of flour, about
three inches in circumference, which surrounds
them like a magic circle, and consists of the minute dust-like spores shed by the fungus that has
attacked them. When more attentively examined, however, the abdomen is seen to be much
swollen, the rings composing it being separated
from each other by inter-spaces, occupied with
white prominent zones of vegetable growth. The
body is a mere empty shell, reduced by the
slightest touch to a dry friable powder, and lined
with a thin, white, felt-like layer of mycelium, the
entire viscera and all the juices being consumed

by the voracious fungus. The dung-fly in certain districts has been almost annihilated by it. This disease has been long familiar to naturalists, but owing to the imperfection of their microscopes, its real nature was not ascertained until a comparitively recent period. It was first accurately determined by De Geer about the end of last century ; and a minute and graphic description is given of it by Goethe, who suffered nothing worthy of notice, however minute, to escape his observation. This, and all other vegetable parasites attacking insects, seem to be one of the powerful and efficient checks provided by nature for restraining within due limits the increase of creatures which, owing to their extraordinary fecundity, rapid development, and unbounded rapacity, would otherwise prove a terrible scourge. It is well known indeed that they help to prevent the destruction of forest-trees, which goes on to an enormous extent in North Germany through the ravages of caterpillars.

The insect Sphærias are found in different countries. In Australia, where a gigantic species occurs on an enormous larva frequently found beside the banks of the Murrumbidgee, in North America, and in China, these deadly parasites are developed upon insects of different tribes. They form a favourite medicine in China, where a bundle of the fungi, with the caterpillars attached to them, is placed in the stomach of a duck, which is then roasted and eaten by the patient as a cure for internal complaints. There is a peculiar disease called muscardine, affecting the silk-worm in Syria and China, before they have woven their cocoons, which sometimes proves fatal to thousands of these delicate creatures. It not unfrequently happens that the silk-grower loses his whole stock of worms from this cause alone. This disease is caused by the mould-like filaments of the *Botrytis bassiana.* These filaments grow with great rapidity within the body of the animal they attack, not only at the expense of its nutritive fluids, but after its death ; all the interior soft tissues appear to be converted into a solid mass of mycelium, from which arise one or more aërial receptacles of the spores. It sometimes happens that the caterpillar is only partially affected by this fungoid growth, or only to such an extent as not to destroy the organs immediately essential to its life, in which case it may pass through its metamorphosis into the imago state, and become a butterfly or a moth, with the lower portion of its body filled with a mass of fungoid substance as above described. According to Pasteur, muscardine is now very rare, but of late years a far more formidable scourge, called *Pebrine,* on account of the black spots which it produces on the skin, has appeared among silk-worms. The mortality arising from this cause has severely crippled the silk industries in the south of France. In 1858 the silk crop was reduced to a third of its former amount ; and up till the last year or two it has never attained half the yield of 1853. This devastating cholera-like pebrine is caused by the growth and multiplication of a microscopic fungus, called by Lebert *Panhisto-phyton*, which appears as a multitude of extremely minute cylindrical corpuscles abounding in every tissue and organ of the silk-worm, and even passing into the undeveloped eggs of the female moth. The further development of this obscure organism has not been recorded.

But it is not only insects, and other creatures of inferior organization in the larva state, that are thus subject to the attacks of parasitic fungi. They even enter the water—an element in which they are seldom found, and where they always refuse to develop themselves normally—and prey upon frogs, fishes, and other tenants of the deep. The *Achlya prolifera* is one of the most remarkable of these fungi. It is supposed to be only an aquatic form of the common fly-fungus or muscardine already described. Every one who has kept gold-fishes must be familiar with this great enemy of his favourites. It consists of numerous trans-

parent threads of extreme fineness, packed together as closely as the pile of velvet, adhering to
the surface of the fishes, and covering them as it
were with a whitish slime. This appearance is
generally regarded as a kind of decay or consumption in the animals themselves, and not as an external clothing of parasitic plants. It is, however,
a true vegetable growth, as is evident when it is
placed under the microscope, for the unassisted
eye can perceive nothing of its true structure ; each
filament being terminated by a pear-shaped ball,
about the $\frac{1}{1200}$th of an inch in diameter, and consisting of a single cell filled with a mucilaginous
fluid, in which float the reproductive granules.
These granules move in currents running in various directions, like the circulation of cell-contents
in the hairs of Tradescantia. As they advance to
maturity, the mucilage disappears, and the motion
of the granules becomes more rapid and violent, till
ultimately they burst their way through the cell, and
are transferred to the water, there to perform their
circle of being, and to give birth to new granules.
All this takes place with such rapidity that one
observer has remarked that an hour or two suf-
fices for the complete development and escape of
the spores ; so that we need not wonder when we
are told that, once established, the *Achlya prolifera*
will often complete the destruction of a healthy
gold-fish in less than twelve hours.

The existence of a flora within the bodies of
living animals is one of the most extraordinary
facts which the microscope has revealed to us.
A curious plant is occasionally met with in the
frothy vomit ejected in severe cases of stomach
disease. It resembles a number of minute wool-
packs strung together, and is known as the *Sarcina ventriculi* of Goodsir. It is found in the
human kidney and lung, as well as in the stomach.
Upwards of ten other varieties of entophytes have
been discovered in various parts of the human
body, and described and figured by M. Robin in
his magnificent work, ' Des Végétaux qui croissent
sur l'homme et sur les animaux vivants.' Man,
however, is less infested by entophytes than other
animals, on account of the cooking process to
which his food is subjected, which effectually destroys the germs of parasites, and his high degree
of organic activity, which is unfavourable to their
development. Animals of feeble vitality and sluggish habits, using solid innutritious food difficult
of assimilation, and therefore remaining long in
the alimentary canal ; or animals swallowing their
food in large morsels, to which the germs of plants
may adhere, are rarely, if ever, free from these
entoparasitic plants, which, however, when few in
number, or not of excessive size, are quite harmless. They are found principally in those portions

of the body which are easy of access from without, such as the stomach and intestinal canal, and where, of course, all the indispensable conditions for the maintenance and reproduction of such life exist. Dr. Joseph Leidy, to whom we are in- debted for much new and interesting information· upon this obscure subject, found the most exten- sive entoparasitic flora with wonderful uniformity within the intestinal canal of a species of myriapod, and a species of beetle living in decaying stumps of trees. The vegetable forms he discovered in this singular situation are exceedingly curious, and notwithstanding their very subsidiary position as parasites, display as high a degree of organization as any of the larger confervæ which inhabit our streams. They consist, in almost every case, of yellowish or colourless transparent tubes, varying from half a line to two or three lines in length, attached to their growing-place by broad disks, and proceeding in a straight or gently flexuose curved line to the free extremity. They are filled with exceedingly minute, faintly yellowish, oil-like granules, enveloped by much larger globules, ar- ranged like a string of beads along the whole interior. Exceedingly minute and obscure al- though these organisms are, they present many beautiful instances of means adapted to the end in view in their form. They are generally fixed upon the mucous membrane of the intestinal canal in which they grow ; and from their delicate structure they must be more or less constantly liable to rupture in the peristaltic movements of the bowels, and the passage onwards of the food ; but from the spiral arrangement of one species, and the sigmoid flexure of another, these graceful filiform plants may be elongated or stretched onwards for a considerable distance without danger of being torn. The economy of these plants is altogether peculiar ; existing as they do under circumstances totally different from those in which all other plants are found, and in this respect strikingly illustrative of the wonderful capacities of vegetable life. That vegetable life can exist within animal life is an extraordinary circumstance ; but a circumstance still more extraordinary in connexion with their history is, that many of these entophytes are developed upon entozoa, or animals within animals, and are in their turn the seat of other parasitic entophytes more minute, while even on these parasitic entophytes themselves are produced still more minute forms of vegetation. We have thus within the microscopic compass of a beetle's body an epitome of what takes place on a large scale throughout the world of sense and sight— life supported by life to the third and the fourth degree. These parasite and parasitic-parasite

entophytes sometimes grow with such luxuriance
that they completely cover the original plants, and
hide their forms from view ; while at other times
they are confined to the extremity of the branches,
or form concentric circles around their base and
middle. They consist of simple or branched filaments, sometimes aggregated into thick radiating
tufts, like a thick tassel held upwards, each thread
or filament measuring from the $\frac{1}{30000}$th to the
$\frac{1}{50}$th of an inch in length, and from the $\frac{1}{30000}$th
to the $\frac{1}{250000}$th of an inch in diameter. Some of
these seem to be articulated, and to contain spores
or germs between the joints ; but owing to their
exceedingly minute size, the power of the microscope in its present condition is not sufficient to
resolve them into their component parts, and of
their processes of growth and propagation we
know absolutely nothing. Dr. Leidy included all
those obscure organisms under the confervoid
algæ, and gave them specific names, such as
Cladophyton, Enterobryus, Arthromitus, under the
impression that they were independent plants.
But modern research has proved them to be the
mycelia or immature forms of various fungi. The
sarcina is beyond doubt merely an algal condition
of the common mould, produced and retained
in that state in the stomach by the special
food which it meets with there, and which it

finds in no other locality, but reverting to its original form when the supply of this peculiar pabulum is exhausted. And as if to establish this conclusion beyond question, it has since been found in precisely the same form as in the stomach, in cases of parasitic skin disease.

Professor Lister has demonstrated that the lamentable mortality which so frequently attends surgical operations, and the deadly consequences of wounds and injuries in great hospitals, are owing to the development and multiplication of fungoid growth. Several French surgeons narrate cases in which, on removing bandages from ulcerated and mucous surfaces, they have found them covered with a collection of white flocculent filaments, evidently the undeveloped spawn or mycelium of some fungus. The connexion between erysipelas and hospital gangrene and the theory of fungoid germs, ought therefore to lead to the most scrupulous cleanliness in conducting surgical operations. The surgeon who will save most lives will be he who adopts the antiseptic method of treatment, and carefully guards against the introduction of fungoid germs by the thorough purification of his hands, instruments, and all the appliances he uses. Besides the mycodermata and mortification of wounds, several other diseases incident to man are of fungoid origin. There is a curious endemic

z

disease called *Plica polonica*, which occurs in Poland
and the adjoining countries, said to be of Asiatic
origin, and to have first appeared in Europe in
the thirteenth century, in which the hairs get
swollen, matted together, and become endowed
individually with the most exquisitely painful sensibility. This fearful disease is caused by the growth
of a fungus upon the head ; the compact mass
of hair being sprinkled over, as with flour, with
its germs. The allied disease, known by the name
of chin-welk or mentagra, affecting the beard of
men ; the sordes on the teeth, occurring in persons
affected with low typhoid fever ; the aphthæ or
thrush, as the white spots like curdled milk which
cover the mucous membrane of the mouth and
palate of infants are called ; the disease called
scald-head and ring-worm, so frequent on the
heads of children, and so highly infectious ; the
pityriasis, or dandriff, which produces pale-brown
circular spots on the neck and breast, accompanied
by abundant desquamation of dry branny scales,
constantly renewed ; the yaws, so prevalent in the
West Indies, and in some parts of Africa ; the
elephantiasis, which so horribly disfigures the
Egyptians ; the ichthyosis or fish-skin of the
East ; the pellagra of the plains of Lombardy and
Northern Italy ; the formidable disease known as
the fungus-foot of India, which is very common

among the farmers or ryots—disorganizing in many
cases the structure of the whole foot, and occasioning much suffering;—all these diseases are either
primarily produced or invariably accompanied by
some form or other of fungoid growth. The vegetable vesicles or aggregations of small rounded
cells found in all of them, have been carefully removed and placed in saccharine matter, on fruits
or in syrup, in favourable circumstances, supplied
with the requisite conditions of warmth and moisture, and attentively watched, when, in the course
of a few days, they were all found to develop
themselves into some species or other of the common mould of our cupboards. These experiments
render it extremely probable that there is no
fungus found infesting any part of the human body,
or any part of the economy of other animals, however different or abnormal the appearance it may
present, which is not referrible to the ubiquitous
mould family.

The plants known under the common name of
mould are not only universally distributed where
fungi are at all capable of growing, but are also
remarkably indifferent as to their selection of
habitats, scarce anything escaping them, assuming
different appearances in different situations, some
of which are exceedingly puzzling to the botanist.
Usually they are found on pots of jam, on cheese,

on decaying succulent fruits, on bread when kept too long in a warm and damp situation, on old shoes laid aside, on clothes and other articles of common wear. There are several genera and numerous species included under the one popular name. Of these the most familiar is the *blue mould*, belonging to the genus *Aspergillus* (Fig. 36), so called from the resemblance of its fructification to the aspergillus or brush used for sprinkling holy water in Roman Catholic

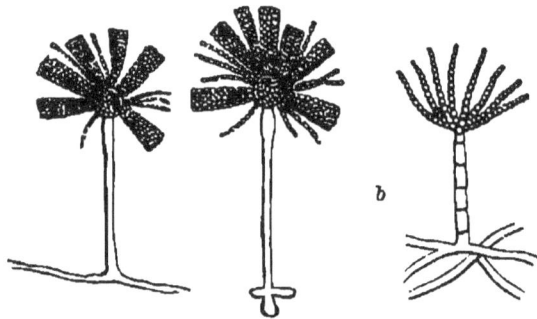

FIG. 36.—(*a*) ASPERGILLUS GLAUCUS.　(*b*) ASPERGILLUS CANDIDUS.

countries. This species is of very frequent occurrence on decaying substances of all kinds, and gives a white and downy, or a blue-grey and powdery aspect to the objects on which it grows. Another exceedingly common species is the *green mould*, belonging to the genus Penicillium (Fig. 37). It presents a close resemblance to the former species, but differs in this respect, that its spore-bearing stem divides into numerous

branches like a miniature tree, bearing spores not
in regular rows, but like leaves or fruit in irregular

FIG. 37.—PENICILLIUM CRUSTACEUM.

clusters on each branch, whereas the stem of the
aspergillus is unbranched, and bears on its summit many rows of spores, which are placed in

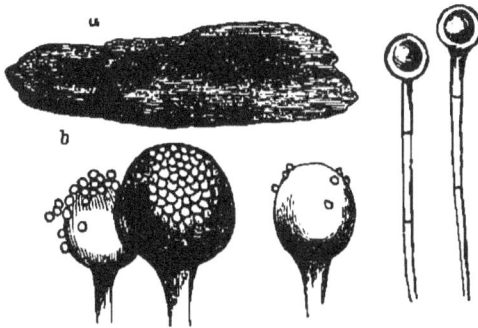

FIG. 38.—MUCOR MUCEDO.

(*a*) Natural size.   (*b*) Highly magnified.

linear order like necklaces, and joined to the stem
like a bundle of hairs on a brush. Another familiar kind of mould is the *Mucor mucedo* (Fig. 38),
or *yellow mould*, also extremely common. It dif-

fers from the two preceding kinds in having its spores, instead of being exposed naked to the air like them, enclosed in a rounded membranous case, bursting irregularly as the spores arrive at maturity, which then present themselves like so many dusty particles congregated round a central nucleus. Being so minute, the slightest touch, or the gentlest breath of air, is sufficient to scatter them in thousands, and thus they increase with amazing rapidity. The trivial names of blue, green, and yellow mould, it may be remarked, are of no specific significance, as all these colours are common to the species in the different genera, and occur even in the same species in various stages of its growth. In fact, it is by their different fructification under the microscope alone that the different genera can be recognised, as their mycelium or spawn is precisely similar, and to the naked eye the appearance they present on the different substances which they affect is identical.

But perhaps the most extraordinary and abnormal forms of mould are those which it assumes in liquids. Fungi, as a class, are confined to solid substances ; but there are very few fluids containing saccharine matter in which this all-pervading mould does not occur. Wine, cider, tinctures, syrups, vinegar, catsup, not unfrequently become mothery, that is, present the appearance of fibres

or flocculent threads running through them.
Every one is familiar with the tough mass that is
so often brought up on the point of the pen from
the ink-holder. This flocculent matter is the undeveloped mycelium of the green or blue mould
While growing in decomposing liquids, it loses all
resemblance to the same plant when growing on
decaying fruits and dead organic matter exposed
to the air, and becomes a soft, slimy, and somewhat gelatinous body, such as is often found in the
bottoms of empty wine-bottles. This slimy mass
is no other than the famous vinegar plant, which
a few years ago aroused the attention of domestic
circles and scientific bodies, and was extensively
diffused as a useful article in the manufacture of
vinegar in private families. The report, circulated
at the time, of its being an importation from India
or South America, has thus been found destitute
of foundation ; for whatever may have been the
history of the first or individual specimens, and
though the growth of the plant might go on more
rapidly in a warm than in a temperate climate, yet
it is evidently a genuine native production, capable
of being originated and multiplied indefinitely in
this country. This extraordinary substance, familiar, no doubt, to many of my readers, may be described as a tough, gelatinous mass of a pale
brownish colour, bearing a close resemblance to a

piece of boiled tripe. It is usually placed in a small jar containing a solution of sugar or treacle; and after being allowed to remain in a warm situation for a month or six weeks, the solution is converted into vinegar, this change being due to a kind of fermentation caused by the plant. If the vinegar plant be left however in the liquid after it has transformed the sugar, it oxidizes the vinegar, and leaves the housewife who has neglected to remove it at the right time only dirty water for her pains. The same substance, it may be remarked, exists as a delicate diffused organic film when other modes of vinegar-making are adopted. It covers the beech or deal shavings, dried and soaked in strong vinegar, over which German manufacturers cause an alcoholic mixture to trickle slowly when they wish to promote its acetification. After a few days, when the vinegar-plant has had time to grow, the process advances with great rapidity. The so-called vinegar-plant is only an abnormal form of the common *Penicillium crustaceum* or blue mould. In fact it is merely the spawn or mycelium of that plant, increased to an extraordinary extent and closely interlaced together, owing to the absence of the usual spore-bearing stalks, which, as already remarked, are never formed in fungi growing in fluids. Whenever the vinegar is allowed to evaporate, and the mycelium in consequence becomes free from saturation, it produces the usual fructification, and presents the common appearance of mould. Other fungi besides the blue mould may assume the same remarkable form when placed under similar conditions, and all of them may have the power of producing vinegar. Indeed, it need not be a matter of surprise, that fungi should assume such extraordinary appearances when prevented from developing their usual organs of fructification, for do we not find even among flowering plants, which are not nearly so plastic, or so susceptible to external influences, very singular changes effected in their structure and conformation by being kept in a barren and undeveloped state? The tree mignonette is a familiar instance of the change effected in the structure of an annual plant by being kept from flowering during the natural period, and placed in favourable circumstances ; and still more surprising illustrations of enormous development of the vegetative at the expense of the reproductive system will occur to the florist and botanist. It is worthy of remark that the vinegar-plant, when well supplied with food in an acetous solution, divides at a certain stage of its growth into two distinct layers, which in course of time will again increase in size and divide, and so on ; each layer being capable of removal to a separate jar for the production of vinegar. This remarkable mode of propagation by dividing into separate laminæ, which has been taken advantage of in spreading specimens of the plant, resembles the separation of buds in the medusæ, and the meristmatic mode of division by which the diatoms, and many others of the lowest class of algæ, extend themselves indefinitely ;—thus showing the close and intimate connexion subsisting between plants that in other circumstances are widely different, when placed under similar conditions.

ᴵ A still more striking form of the protean mould under consideration, is that which occurs in the fermenting of yeast and other substances. Indeed, some botanists are of opinion that the vinegar-plant is the vegetative growth of the mould taking place at low or ordinary temperatures in highly saccharine liquids ; while the yeast-plant is formed in the more rapid fermentation taking place at more elevated temperatures. It may surprise many to be told that yeast is merely an undeveloped condition of the common mould which they see on their bread and cheese. Fermentation is in one sense a chemical process, forming the first step towards dissolution, or that re-arrangement of old elements which is necessary to form new compounds ; but, strange to say, the action is also vegetative. The whole mass of fermenting matter gradually assumes the condition of active vegetative growth. The germs of the mould, which had been incorporated in the material, begin to live and expand, each bearing a distinct plant, giving rise either by gemmation or nucleation to new plants indefinitely, until the entire fermenting principle is exhausted. The form which the *Torula cervisiæ* or yeast-plant assumes is that of a number of small spherical vesicles, which when mixed with the wort of beer speedily produce one or two gemmæ or buds. These buds grow to the size of the parent cells, and produce new buds ; so that in several hours a string of cells is formed like a necklace of beads. By the time that five or six vesicles are strung together, the fermentation is sufficiently advanced, and the manufacturer checks it. The vegetation is then suspended, and the groups of vesicles separate into individuals, the mass of which thus constitutes the yeast. The cells of the yeast-plant are globular at first, but they gradually change, while the fermenting principle is being used up, into the oval form ; when the sugar is still more exhausted, they become linear and filamentous, advancing to the primary stage of mycelium ; until finally, when the whole fermenting matter is absorbed and evaporated, they develop into the normal crust and organs of fructification of the common Penicillium or blue mould.

Yeast is very tenacious of its vitality, its power
of growth not being destroyed either by a heat
almost equal to that of boiling water, or by a cold
of 76° below zero, or by being completely dried up.

For what purpose, it may be asked, were plants
so excessively numerous, and so universally distributed, created ; for to many individuals they
are such objects of prejudice and disgust, that
their real importance as useful productions is little
appreciated? We do not know indeed *all* the wise
purposes which He who created nothing in vain
intended them to serve in the economy of nature ;
but we are acquainted with some of them, and
these are so obvious and important, and reveal
such striking examples of adaptation of means to
ends, that we cannot but lament that such ignorance and prejudice regarding them should exist in
this country. There is no elementary and self-
subsistent organic matter in nature, as Buffon
erroneously taught ; and the health and wellbeing
of man himself may more or less immediately depend upon the offices which these despised productions perform. We have seen the effect of
fungi in the two great processes of fermentation
and putrefaction, which are of such vast importance
in the economy of nature and art, and whose
true cause and nature are now only being understood. Appearing as they do in those months of

the year when the flowers are fading, the leaves
falling, and all nature yielding herself up as the
passive victim of decay and death, fungi are obviously intended to remove those decomposing
tissues which would otherwise pour volumes of
noxious vapour into the atmosphere, and render it
unfit to support life ; to call back into the great
vortex of existence those fugitive particles of effete
matter which had served their appointed purpose
in one form of organization, and were fast hastening down, by a process of decomposition, to join
the atoms of the inorganic world of chaos and
death.   Every decaying leaf of the wood and the
field has its own fungoid parasite, which gradually
reduces it to a state fitted to minister to the
necessities of next year's vegetation ; and thus,
through the agency of little insignificant patches
of mouldy, rusty tissue, the carrion in the sun converts itself into trees and flowers.

In the economy of man, fungi have been applied
to many useful purposes.   A few are endowed
with valuable medicinal properties, and still hold
their ground, notwithstanding the vast improve-
ment effected in the nature and choice of drugs in
recent times.   From their chemical constituents,
the medical uses of the fungi are probably of far
greater importance than their present very limited
application might lead us to suppose ; and in all

likelihood, if they were more studied, many of the active species might afford valuable remedies. As it is, however, one species at least is a highly powerful medicine. The ergot of rye is an important article in the Materia Medica, as it has been found capable of exerting a specific action upon the womb, and is administered in small doses in certain extreme cases. This remedy has been principally used in America, although of late it has been successfully employed in France and in this country. Dufresnoy is said to have used *Agaricus emeticus* with success in the early stage of consumption ; and the sweet-scented Polyporus has been much vaunted for its surprising effects in the treatment of the same disease, but it has now fallen entirely into disuse. Under the name of *Lycoperdon nuts* or *Hart's truffles*, the *Elaphomyces granulatus* used to be employed on account of its supposed medicinal properties, and it may still be seen occasionally in herbalists' shops in country districts. A species of Polyporus growing upon the birch is used, when dried and pounded, as an ingredient in snuff, by the Ostyacks on the Obi. *Lysurus mokusin* is used by the Chinese as a remedy in gangrenous ulcers, but its virtues are probably fabulous. In Lapland, the common amadou (*Polyporus fomentarius*), when beaten out into thin pieces, is employed to remove pain by simply laying a piece of it on the part affected. Like the soft contents of puff-balls, it is used occasionally to stanch blood in wounds, and as a sovereign remedy for a cut finger. When steeped in saltpetre, and cut into thin slices, it forms most excellent tinder, and is employed in the form of *fusees* by smokers in Germany and England. In Lapland, it is considered an indispensable article in domestic economy, Linnæus relating that he saw it hung up for various purposes on the walls of every cottage he entered. The *Hirneola* or *Jew's-ear* had at one time a reputation for the cure of sore-throats, and also as a topical astringent. Owing to its power of absorbing water like a sponge it is sometimes used as a medium for the application of eye-water to sore eyes. There is a very curious fungus found in the interior of old deserted white ant-hills in the Coimbatore and Malabar districts of Southern India, which is gathered by the natives as a specific in cholera and a variety of other diseases, and highly prized. Its Indian name is *Puttu Mango*, or White Ant-hill Mango, while to botanists it is known as the *Sclerotium stipitatum*. During the monsoons it attains the size of an orange. It grows in clusters of six or more, hanging each by a separate stalk from the roof of the outer passage of the ant-hills, or simply overlying the floors of the cells with-

out stalks. Its form is very varied, being generally oblong and irregularly round. The external
rind is black and slightly wrinkled ; while the
interior is white and pithy, and is compared by
the natives to the kernel of a tender cocoa-nut.
It is tasteless and inodorous. By snake-charmers
it is collected and sold as the favourite food with
which they supply their dancing-serpents. Some
of the villagers attribute to it poisonous properties.
Strange to say, the natives of West Africa worship
the magnificent *Polyporus sacer ;* while another of
their fetiches is the *Yatum condenado,* or *accursed
devil,* a species of Lycoperdon.

Many of the fungi are possessed of highly poisonous properties, and serious, and even fatal,
accidents occur occasionally in this country, and
more frequently in France, from an incautious use
of them. Sometimes this arises from confounding
the edible with the poisonous species ; but even
the edible kinds, owing to peculiar idiosyncrasy in
some people, act always as poisons, and there is
reason to believe that the best and safest mushrooms, if taken in considerable quantity for any
length of time, induce in many individuals a habit
of body which may be pronounced a poisoned one.
Pliny relates that Annæus Serenus, the prefect of
Nero's guard, with his tribunes and centurions,
accidentally met their death by eating some poi-

sonous fungus. Seneca, the friend of Serenus, had
an intense aversion to the whole tribe. He says :
' Do you think that those *boleti*—a pleasant poison
—albeit they hurt not now, concoct within them
no hidden mischief?' Nero, with villanous irony,
called them the 'food of the gods.' Upon what
the poisonous. properties of fungi depend is not
known. Two active substances have been recognised in them. When distilled with water they
yield a fugacious acrid principle, dispelled in the
act of drying, or by immersion in acids, alkalies, or
alcohol. When extracted by water and alcohol,
a brown, solid substance called amanitine is obtained, which is more fixed, and resists such processes. The specific action of these two constituents of the poisonous fungi upon the human frame
has not as yet been investigated. They sometimes act like narcotics, producing comatose and
other affections of the nervous system, and at
other times their action is of an irritant nature,
more approaching that of arsenic. Some act as
anæsthetics, giving complete insensibility to pain ;
while, unlike chloroform and ether, the individual
under their influence remains conscious all the
time. The common puff-ball deprives the patient
of speech, motion, and sensibility to pain, while he
is still conscious of everything that happens around
him ; thus realizing that nightmare of our dreams

in which we lie stretched on the funeral bier, sensible to the weeping of friends, aware of the last screw being fixed in the coffin, and the last clod clapped down upon us in the churchyard, and are yet unable to move a hand or a lip for our own deliverance. When slowly burnt, this fungus has long been employed for stupifying bees, and thus robbing their hives of the honey with impunity. Experiments, with the same species, have also recently been made on dogs, cats, and rabbits, and similar effects have invariably been found to ensue. When the fumes of the burning fungus are slowly inhaled, they gradually produce all the symptoms of intoxication, followed first by drowsiness, and then by perfect insensibility to pain, terminating, if the inhalation be continued, in vomiting, convulsions, and ultimately in death.

The qualities of fungi seem to vary with the climate in which they have grown ; for many species which in this country are considered highly poisonous, on account of their intensely acrid qualities, and avoided as such, are eaten with impunity on the Continent. Mr. Berkeley mentions his having been informed by a gentleman of great acuteness and observation, that in some town of Poland, where he was detained as a prisoner, he amused himself with collecting and drying the various fungi that grew within its walls, amongst which were many reputed dangerous, and that to his great surprise his whole collection was devoured by the soldiers. It is well known, indeed, that even the esculent fungi of this country are not always safe to eat ; the qualities sometimes varying very considerably according to the nature of the situations in which they occur. The common edible mushroom of this country (*Agaricus campestris*), the pratiola of Italy, has sometimes proved fatal on the Continent, and by the inspectors is invariably excluded from the Italian markets as most pernicious. Owing to this circumstance it has given rise to the Italian's curse, ' May he die of a pratiola.' The ancient Romans, however, were wiser in their choice, for Horace lays down the rule that meadow mushrooms are the best of the whole tribe. The most useful and innocent species become poisonous when growing in damp, dark localities, such as old decaying forests, and cellars where there is little circulation of pure air. At different stages of their growth, also, fungi manifest different qualities ; a few hours being sufficient in some instances to change a nutritious into a poisonous substance. Their fitness for the table seems also to depend much upon the rapidity of their growth; those which grow slowly being of tougher texture and less delicate flavour. A warm sun after heavy rain brings them out in greatest perfection. All agarics are more wholesome fresh than stale, and should therefore be prepared for the table as soon as possible after being collected.

The intoxicating Siberian fungus or Fly Agaric (*Agaricus muscarius*, Fig. 39), so called because a decoction of it used to be employed as a fly poison, may be adduced as an illustration of the remarkable effects produced by some species of fungi, when growing in foreign countries. We have no experience as yet, in this part of Europe, of any effects so extraordinary being produced by any of our native fungi, or even by the same species when growing in the British woods. It is acknowledged to be one of the most poisonous species in this country ; but it does not exhibit its curious properties to the same extent here, as it does beyond the Ural Mountains. In European Russia, and Siberia, this fungus is to the inhabitants what opium and hemp are to the natives of India and China ; cocoa to the Peru-

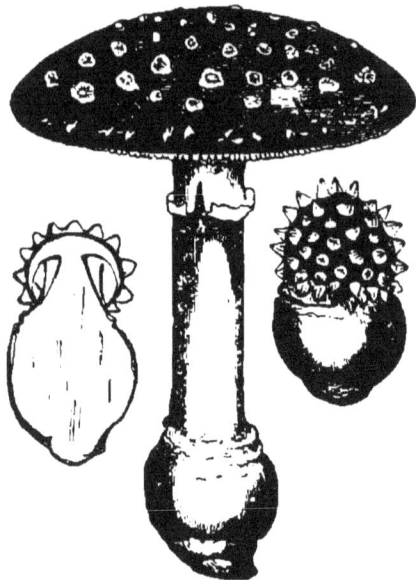

FIG. 39.—AGARICUS MUSCARIUS.

vians ; and tobacco to the inhabitants of Europe
and North America. The craving for narcotic
indulgences, so natural to the human race, has,
among the Kamtschatkans and Koriacs, found its
gratification in an object so low in the scale of
nature as a common toadstool. These races are
so dreadfully degraded, that they personify this
fungus under the name of Mocho.Moro, as one of
their household deities, somewhat like the Goddess
Siva of the Hindu Thugs. If they are urged by
its effects to commit suicide, murder, or some
other dreadful crime, they pretend to obey its
commands ; and, to qualify themselves for pre-
meditated assassination, they have recourse to
additional doses of this intoxicating product of
decay and corruption.

This plant, around which such a lurid interest
gathers on account of its debasing associations, is
by no means rare in this country ; in fact, it appears to be very generally distributed throughout
the whole of the temperate zone. In the Highlands of Scotland, and the northern districts of
England, it is very common and abundant, particularly in woods of fir and birch, where its tall
white stem and rich orange scarlet cap, studded
with white scaly warts, frequently of portly dimensions, form a beautiful contrast to the green carpet
of moss from which it springs, and the draperies

of green foliage that overshadow it. It is exceedingly abundant in some parts of Kamtschatka and
the northern districts of Siberia ; the ground, in
nearly every wood and thicket, being almost concealed by its scarlet sheen. By the natives it is
collected during the brief summer months, which
in that climate are intensely hot. Sometimes it is
plucked and hung up in the air outside their dwellings to dry, and sometimes it is allowed to wither
and die untouched in the place where it grows, in
which case its narcotic properties are better preserved than when it is gathered and artificially
dried. When steeped in the expressed juice of
the native whortleberry, it forms a very strong intoxicating kind of wine, which is much relished.
But the more common way of using the fungus is
to roll it up like a bolus, and swallow it without
chewing, which, it is said, would disorder the
stomach. Dr. Greville gives some curious particulars regarding this fungus in the fourth volume
of the *Wernerian Transactions.* He says : 'One
large or two small fungi are a common dose to
produce a pleasant intoxication for a whole day,
particularly if water be drunk after it, which aug-
ments the narcotic action. The desired effect
comes on from one to two hours after taking the
fungus. Giddiness and drunkenness follow in the
same manner as from wine or spirits. Cheerful-

ness is first produced, the face becomes flushed, involuntary words and actions follow, and sometimes at last entire loss of consciousness. It renders some remarkably active, and proves highly stimulant to muscular exertion. By too large a dose violent spasmodic effects are produced. So exciting is it to the nervous system of some individuals, as to produce effects which are very ludicrous. A talkative person cannot keep silence or secrets, one fond of music is perpetually singing, and if a person under its influence wishes to step over a straw or small stick, he takes a stride or a jump sufficient to clear the trunk of a tree.' The intoxication produced by this fungus sometimes amounts to absolute delirium, and not unfrequently terminates in convulsions, coma, and death ; and it is a most remarkable fact that it communicates its narcotic properties to the fluids of the debauchee, which, in consequence, are carefully preserved and eagerly consumed during the winter months, when the season of the plant is over, and the stock of dried specimens is exhausted. Thus a whole village is intoxicated through the medium of one man, and a few fungi serve to prolong these disgusting orgies for many days at a time. The very same erroneous impressions as to size and distance produced by this plant, are also created by the haschisch and majoon of India, and are frequently

noticed among idiots and lunatics. It is not im-
probable that many poor half-demented creatures,
particularly if old and ugly, have suffered martyrdom at the stake during the witch-mania of Scotland owing to this natural defect; inability to
step over a straw being considered the conclusive
test of familiarity with evil spirits. It may be
mentioned that the Fly Agaric is historically interesting as having caused the death of the Czar
Alexis by eating it; for notwithstanding its highly
dangerous properties it is sometimes cooked and
eaten in Russia.

Passing from the consideration of the noxious
properties of fungi, they exhibit themselves to us
now in a more interesting and pleasing aspect as
edible substances. In common with other classes
of plants which have the reputation of being
poisonous, and yet contain several esculent species
invaluable to man, the fungi, although considered
as a class dangerous and unwholesome, yet yield
in many instances a large and varied supply of
palatable and nutritious food. No country is perhaps richer in edible fungi than Great Britain; but
such is the extent of wilful ignorance and silly prejudice regarding them, arising from their cold,
moist, clammy nature, and the disagreeable situations in which they often grow, that this savoury
and important food is year after year allowed to

perish ungathered in the woods and fields. Dr.
Badham, in his excellent work on the esculent
fungi of this country, remarks regarding this culp-
able neglect: 'I have myself witnessed whole
hundredweights of rich wholesome food rotting
under trees; woods teeming with food, and not
one hand to gather it; and this perhaps in the
midst of potato-blight, poverty, and all manner of
privations, and public prayers against imminent
famine. I have indeed been grieved to see pounds
innumerable of extempore beefsteaks growing on
our oaks in the shape of *Fistulina hepatica ; Agaricus fusipes* to pickle in clusters under them ; puff-
balls, which some of our friends have not inaptly
compared to sweet-bread, for the rich delicacy of
their unassisted flavour; *Hydna* as good as oysters,
which they somewhat resemble in taste ; *Agaricus
deliciosus*, reminding us of tender lamb-kidneys; the
beautiful yellow Chantarelle, that *kalon kagathon*
of diet, growing by the bushel, and no basket but
our own to pick them up ; the sweet nutty-flav-
oured *Boletus*, in vain calling himself *edulis* where
there are none to believe him ; the dainty Orcella,
the *Agaricus heterophyllus*, which tastes like the
craw-fish when grilled ; the *Agaricus ruber* and
Agaricus virescens, to cook in any way, and equally
good in all ; these are among the most conspicuous
of the edible fungi.'

There are at least forty kinds of esculent fungi in Great Britain which may be safely used at table, and are as good, if not better, than the common mushroom, which appears to be the only species whose merits are at all appreciated. *Agaricus gambosus* or St. George's mushroom, so called from its usually appearing in this country as early as St. George's day—about the beginning of May— though generally rejected by housekeepers in the country as unwholesome, is frequently sold in London, under the name of Whitecaps. The flavour, however, is far inferior to that of the common mushroom ; its smell is strong and un-pleasant, and it is little fit for making ketchup, having but a small quantity of juice, and that not of a good colour. It grows to an enormous size, frequently attaining forty inches in circumference, and weighing many pounds. It is easily known by its white pileus and gills, slightly stained with yellow when bruised. In France it is known by its white colour as the *Boule-de-neige.* In France and Italy it is so highly prized that when dried it will realize from twelve to fifteen shillings per pound. There is another fungus frequently sold in Covent Garden market under the name of Blewitts, whose taste is very agreeable. This is the *Agaricus personatus*, occurring abundantly in old pastures during the winter months, and often growing gregariously in large rings. It is easily known by its pale bistre or purple-lilac colour, and its rather overpowering odour. Every one is familiar with the common champignon or Scotch bonnets, which form those sour ringlets in the grassy meadows popularly called fairy rings, strangely attributed by some authors to the effects of electricity, and by others, more poetically and quite as truly, to the fairies as the traces of their moonlight revels. This curious fungus, the *Marasmius oreades* of botanists, though tough and strongly tasted, is sometimes used as an article of food in this country, but too frequently very different and poisonous fungi are gathered under the name. It is almost always gregarious, growing in a centrifugal manner, increasing its circle year by year, while the individuals in the centre decay, and impart by their decay to the grass at the edge a more vivid green than that of the rest of the meadow. Among the recent species generally admitted as wholesome may be mentioned the Parasol Agaric (*Agaricus procerus*, Fig. 29). It is known by the thick skin of its cap, which breaks up into distinct scales. It is greatly esteemed on the Continent, where it is known in Italy as the *bubbola maggiore*, in Spain as the *cogomelos*, in France as the *coulemelle*, and in Germany as the *parasol schwamm*. In almost every rich pasture, and often

in gardens, in autumn, may be found the Maned
Agaric (*Coprinus comatus*). It is easily known by
its graceful form, rising from the ground like a
cylinder with a rounded end. The long silky cap
trembling upon its snowy stem breaks up into
scales, and at the margin splits into threads like
the end of a wig. The gills pass through shades
of pink purple and brown to black ; the last stage
being quickly reached, presaging the immediate
dissolution of the whole plant, which deliquesces
into an inky black fluid. If gathered young it
affords no despicable dish. Dr. Bell calls it the
Agaric of civilisation, owing to its being found
about human dwellings. Among the most interesting and beautiful of the mushroom-tribe is the
Amanita Cæsarea, distinguished by its brilliant
red cap, perfectly smooth, and its rich yellow gills.
Unknown in this country, it is the pride of Southern
markets, on account of its delicious taste when
cooked. It has been known from the time of
the Romans, and is the Boletus of the Satirists,
because owing to its superlative qualities it was
often made a vehicle for poison. Indeed, its
specific name is derived from the supposition that
it was the identical species in which Locusta, at
the instigation of Agrippina, conveyed poison into
the stomach of the Roman emperor Claudius, who
was excessively fond of it. In America it is highly
prized. under the name of *Imperials.*

Some of the species mentioned in the paragraph quoted from Dr. Badham are rather suspicious objects of food, and although they may sometimes be taken with impunity, it is best as a general rule to avoid them. The *Agaricus ruber*, for instance, is a remarkably beautiful and tempting-looking fungus, having a rich orange or a rose-red cap and snowy gills, but its taste is hot and acrid like that of the mezereon or the cuckoo-pint. Though excellent for food, if properly prepared, it is pronounced by Trattinick to be very unwholesome in a raw state ; and M. Roques' account of it is even more un- favourable. The same objection applies to the *Lactarius deliciosus*, said by Dr. Badham to re- mind him of tender lamb-kidneys. The odour and taste of this Agaric are agreeable ; but from the account of it given by M. Roques, it would appear that, however delicious, it is not always to be eaten with impunity. These two last-mentioned fungi belong to a very remarkable group of the genus Agaricus, called *Galorrheus*, from the milky juice which every part of them exudes when bruised or ˙ broken. This milk is like that of the Euphorbia or spurge when pierced, and like it too is frequently extremely acrid, causing irritation and slight in- flammation in the parts with which it comes in contact. When dry it forms an unctuous mass, which burns with a brilliant flame. It is generally white, like cow's milk, but in some species is variously coloured, being of a bright orange, turning to a dull green upon exposure in *Lactarius deliciosus.* Like the milk in the laticiferous vessels of the flowering plants, such as lettuce, dandelion, chicory, and celandine, it exhibits singular movements under the microscope. Minute molecules are observed to move about in it with extreme rapidity, exactly like those observable in gamboge mixed with water. These may be phytozoa, being connected in some mysterious manner with the reproduction of the plant. It is sufficient to mention that this singular group of Agarics contains some of the most poisonous and deadly of all fungi, and that all the species are possessed more or less of the same acrid and narcotic properties, to justify caution in the use of the two members of the group quoted by Dr. Badham as esculent, however bland and agreeable they may sometimes be found. With regard to the other species mentioned by this author, they may be used with perfect safety, having stood the test of a pretty long and general experience. The *Boletus edulis* (Fig. 40), common in woods and pastures all summer and autumn, and easily known by its broad, smooth, dark umber cap, and white tubes and fawn-coloured stipe, is a most valuable article of food, resembling in taste the common

mushroom, and having this advantage, that it
abounds in seasons when a mushroom is scarcely
to be found. Mr. Berkeley remarks that it can be
cultivated like the mushroom, but by a much
simpler process, as it is merely necessary to moisten
the ground under oak-trees with water in which a
quantity has been allowed to ferment. The only
precaution requisite is to fence in the portion of
ground destined for its production, as deer and

FIG. 40.—BOLETUS EDULIS, SHOWING TUBES INSTEAD OF GILLS ON THE
UNDER SIDE OF THE CAP.

pigs are very fond of it. This method is said to
be infallible, and is practised in France in the Dé-
partement des Landes. In the Departement of
Gironde great quantities are strung on threads,
and dried for the Parisian market. It is preserved
in the same manner in Russia, and is largely
consumed during the fasts of the Greek Church,
being soaked in water till it becomes soft, and then

cooked. In Lorraine it is a favourite dish, under the name of the Polish mushroom. It is well to mention, however, that the genus Boletus contains perhaps the most poisonous of all the fungi. The *Boletus Satanas* is so called from its deadly properties. It may be known by the underside of its cap, and its stem being of a blood-red colour. The tubes under the pileus of many of the species

FIG. 41.—MORCHELLA ESCULENTA.
Reduced half size.

change on pressure from yellow to blue ; and all those species which exhibit this change, or possess reddish stems or red tubes, should at once be rejected.

Next to the common mushroom, the morel (*Morchella esculenta*, Fig. 41), is everywhere esteemed as a valuable and delightful article of food, and as a condiment to heighten the flavour of ragouts. It is unfortunately by no

means common in this country, being most
abundant in the south of England. It grows
usually in woods, orchards, and cinder-walks
in spring and early summer. Its appearance
is somewhat singular and easily identified,
though like most widely diffused species it
assumes a great variety of forms. It consists
of a hollow stem from one to three inches
high, with its surface ribbed or latticed with
irregular sinuses, surmounted by a round or
conical, hollow, olive-coloured cap about the size
of an egg. Its whole substance is wax-like
and friable. We are informed by Gleditch that
morels grow in the woods of Germany, in the
greatest profusion in those places where charcoal has been made. Hence those who collect
them to sell, receiving a hint how to encourage
their growth, have been accustomed to make
fires in certain spots in the woods, in order to
obtain a more plentiful crop. This strange
method of cultivating morels being, however,
sometimes attended with dreadful consequences,
large woods and plantations being destroyed, the
magistrates interposed and put an end to the
practice. A nearly allied species, called *Helvella
crispa*, is also highly esteemed in some quarters
as an agreeable esculent, though hardly known in
this country. It is a remarkable-looking fungus,

occasionally occurring in woods in autumn. The
stem is from three to five inches high, snowy-
white, irregular, hollow, deeply furrowed, often
full of holes or sinuses like the fluted trunk of
the yarroura or paddle-wood of the Indians. The
cap is deflexed, and commonly divided into curled
or folded lobes which adhere to the stem, but
it is extremely irregular and variable, and
has neither gills nor pores. Its substance is
extremely brittle, the surface being soft like
satin.

The most valuable, however, of all the esculent
fungi is probably the truffle (*Tuber cibarium*, Fig.
42). This curious sub-
terranean puff-ball, for
such it is, is so local and
scarce that it is very
little known except
amongst wealthy and
titled families in this
country, seldom appear-
ing at common tables ; and probably the greater
part of what is sold is imported. It is very rare
in Scotland, but by no means uncommon in some
parts of England, especially where a limestone
soil prevails. It is usually found in beech-woods
and oak plantations growing in clusters or in a
sort of semicircle in the same way as mushrooms,

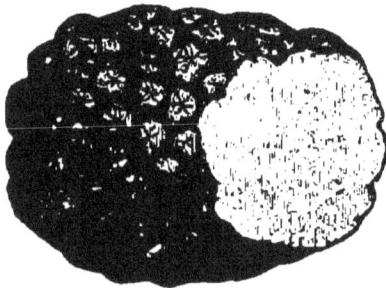

FIG. 42.—TUBER CIBARIUM.

half a foot or a foot beneath the soil. In appearance it is a rounded, rough, irregular nodule like
a potato ; at first white, afterwards black, cracked
like a pine-apple, or a pine-cone, into small
pyramidal or polyhedrous warts. The internal
substance is solid, of a dirty white or pale brown
colour, grained like a nutmeg with darker serpen-
tine lines. The white portions are considered by
botanists to be homologous to the mycelium or
spawn of other fungi, as their structure is decidedly
filamentous ; while the veins are the reproductive
parts, containing in their cellular tissue minute
oval capsules, with two, four, or eight globular,
yellowish, warted sporidia in their interior. This
curious structure, having all the parts of nutrition
and reproduction enclosed internally, instead of
externally as in other fungi, reminds one of the
flower of the fig, which, it is well known, is fixed
upon the inside of the receptacle that constitutes
the fruit. The truffles of Great Britain seldom
exceed three or four ounces in weight ; but in
Italy and Germany they have occasionally been
found weighing eight and even fourteen pounds.
They are received at our tables either fresh, and
roasted like potatoes, or dried and sliced into
ragouts. They are esteemed for their delicious
taste, and are much sought for as a luxury, being
hunted by dogs trained for the purpose. Pigs are

very fond of them, and advantage is taken of their
instinctive knowledge of the spots where they are
found, and their natural propensity to dig them
up, to gather a more plentiful supply than could
be obtained by a chance search.   Nees von Essen-
beck relates an instance of a poor crippled boy
who could detect the hiding-places of truffles with
more certainty even than the best dogs, and thus
earned a comfortable livelihood.   In some parts
of France, truffles are detected by observing a
species of fly which alights on the ground above
the spots where they grow.   They have been
successfully cultivated by Bornholz.   They are
found in dry and light calcareous soil in woods
throughout the whole of Europe, as well as in
Japan, India, Africa, and New Zealand.   In the
warmer parts of Europe, which is the most prolific
region of truffles, all that is needed to procure a
constant supply is simply to sow a few acorns in an
enclosed spot on the calcareous downs.   As soon
as the young oaks have grown a few feet in height
the truffles appear, and a harvest is obtained for
many years successively without further trouble.
There are forty distinct species in Great Britain ;
and France and Italy possess double that number.
Very few of these, however, are used as articles of
food or condiments.   Though some species have
dark and sombre hues, there are many others that

are brightly and variously coloured. One species, the *Hymenogaster carotæcolor*, is of a rich orange hue, and communicates a clear lemon tint to substances with which it comes into contact. Most of the species are dry, but one abounds with a milky juice, and has the peculiar smell of the milky mushrooms. The *Tuber magnatum* or Italian truffle, which, from its high qualities and its great rarity, always fetches a large price in the market, has a smooth outer surface, and looks inside like a slice of pale beet-root. Some species have a strong scent of assafœtida, and are utterly detestable. Modifications of shape, structure, and fructification are very numerous among the truffles. Indeed it is doubtful if they can be regarded as plants with a distinct autonomy and independent existence ; being probably mere hypogæous or underground forms of common aërial fungi, such as mushrooms and Boleti or Pezizas, Helvellas and morels, whose true character and affinities are completely masked by their unusual mode of growth, or perhaps hybernating forms or totally abnormal conditions of mycelium or spawn.

Such is a brief description of the most common and important of the edible fungi known or used in this country. The fatal mistakes which have been sometimes made, by confounding some of them with nearly allied species of a highly

poisonous character, have made them less popular
than they deserve, and increased the national dis-
inclination to the use of any fungus save the common mushroom.   On the Continent, however,
fungi afford not merely a flavouring for a delicate
dish, or a pleasant sauce or pickle, but the staple
food of thousands of the people ; indeed, for
several months in the year, especially in Poland
and Russia, they constitute not only the staple,
but the sole food of the peasantry, and from this
circumstance they are called by enthusiastic
writers 'the manna of the poor.'   To many who
are not reduced by necessity to use them as food,
they form a valuable source of income by collecting them for the market.   Scarcely any of the
four or five hundred species belonging to the
genus Agaricus is rejected by the inhabitants of
northern Europe, with the exception of the
dung and fly Agarics, whose loathsome and
poisonous properties are such as to deter the
most devoted mycophagist from their use.   Even
species which are elsewhere universally avoided
as poisonous, acrid, or disagreeable, are eaten in
these countries with impunity and relish ; their
noxious properties, if not neutralized by soil and
climate, being removed by a process of drying, or
pickling in salt and vinegar.   M. Roques, in his
Histoire des Champignons, gives a very interest-

ing account of a large variety of fungi which
may be used as food. The golmelle of Lorraine
(*Agaricus rubescens*) ; the jozollo and piopparello
of Italy (*Agaricus eburneus* and *pudicus*) ; the
verdette and mouceron of the French (*Agaricus
virescens* and *prunulus*) ; the Nagelschwamm of
Austria (*Agaricus esculentus*) ; the Ziegenbart,
gombas, Brat-bülz and Stockschwamm of Germany (*Clavaria fastigiata* and *coralloides, Boletus*

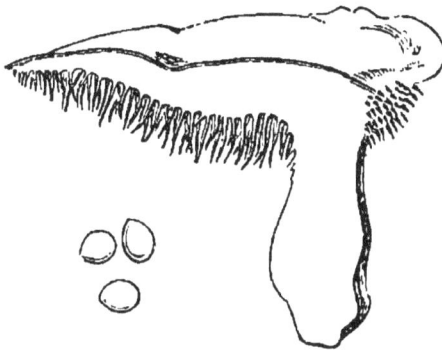

FIG. 43.—HYDNUM REPANDUM, SHOWING SPINES INSTEAD OF GILLS ON THE
UNDER SURFACE OF THE CAP.

bovinus and *Agaricus mutabilis*) ; the Hallimasch
of Vienna (*Agaricus melleus*); the Eurchon or
Barbe de Vache of the Vosges (*Hydnum repandum*, Fig. 43); the Pied de Coq or Gallinole
of France (*Clavaria cinerea*) ; abundantly evince
the great regard entertained on the Continent for
species which, year after year, are suffered to
perish unknown and ungathered in this country.
The common mushroom is consumed in enormous

quantities in Paris, where its flavour is far superior
to ours. All the specimens that appear in the
market are reared in the catacombs. By some
European nations the edible species are eaten raw
and uncooked, as they are considered to be more
palatable and nutritious in their natural state.
Schwaegrichen informs us, that in consequence of
seeing the peasants about Nuremberg eating raw
mushrooms, seasoned with anise and carraway-
seed, along with their black bread, he resolved to
try their effect himself, and that during several
weeks he ate nothing but bread and raw fungi, as
*Boletus edulis, Agaricus campestris, Agaricus pro-
cerus,* etc., and drank nothing but water, when, instead of finding his health affected, he rather experienced an increase of strength. During the
latter part of the American war, when meat was
scarce and dear, fungi, which grow in immense
profusion and variety in America, formed the
principal food of the Southern army. Many
species of fungi have been used for food from
time immemorial in China, whose thrifty inhabitants make the most of the productions of their
native soil, and easily find substitutes among cellular plants when their usual food fails them in a
season of famine. In India and Africa, likewise,
the few edible species that occur have always been
highly esteemed ; our common ketchup, it may be

remarked, being an Indian invention. A kind of fungus called *Mylitta Australis*, which grows on the trunks of trees in Van Diemen's Land, and resembles, when dry, hard compacted lumps of sago, is so frequently used by the aborigines that it is called 'native bread;' while in the wild and desolate island of Tierra del Fuego, the inhabi-- tants subsist, during several months, principally upon a bright-yellow latticed fungus, growing in great abundance on the ever-green beech-trees, and called *Cyttaria Darwinii* after the accomplished naturalist of the 'Beagle,' and the author of *The Origin of Species*. In New Zealand, the gelatinous egg or vólva of a species of Phallus called *Ileodictyon*, is eaten by the natives under the name of *paru watitiri* or thunder-dirt. It has an execrable taste and loathsome smell, in common with the rest of its allies, though its jelly-like consistence would seem to indicate nutritive qualities.

Fungi are to a certain extent capable of artificial propagation, vast quantities of the higher kinds being constantly cultivated for the table. In Italy, a species of Agaric is raised from the grounds of coffee; and a kind of Polyporus, which is greatly relished, is grown simply by singeing the stumps of cob-nut trees, and placing them in a moist, dark cellar. There is a curious production called the fungus-stone, or *Pietra funghaia,*— supposed to be a species of truffle, but in reality nothing more than the spawn or mycelium of *Polyporus tuberaster*, traversing masses of earth which it collects about it in a compact form,—constantly employed for the propagation of that favourite fungus, whose stem and pileus it readily produces when supplied with the requisite conditions of moisture and temperature. The cultivation of the common mushroom is too well known to require comment. Though considered a somewhat precarious crop, it is in the power of almost everybody to grow it, and when carefully conducted it yields a profitable return. This well-known species has almost entirely superseded the wild variety which is now very rarely to be met with in our woods; as is the case with all the animals and plants which man takes under his protection. Mushroom spawn is sold by nurserymen in cakes, and for use is broken into pieces of about two ounces weight. When placed either in a cellar, out-house, or shed, where the covering is effective, in a bed of soil well worked into a compost by the droppings of horses and the parings of their hoofs, and allowed to heat to the temperature of new milk, it is certain to produce a plentiful crop. Microscopic fungi, such as moulds and mildews, may be easily cultivated in our houses by simply sowing their spores on rice paste or on any suitable preparation. The disease of the silkworm and several other epizoic fungi are also readily propagated by inoculation; while the common bunt may be communicated artificially to grains of wheat by rubbing them with the dark powder; and the rust of the garden rose may be made to infect healthy leaves by watering the ground around the bush with a decoction of diseased leaves.

Fungi afford a remarkable illustration of the fact almost universally observed, that agencies which are generally beneficial sometimes prove destructive. While performing their office as the scavengers of nature, these plants sometimes carry their operations too far, and by their rapid increase, and their devastating effects on the fruits of the earth, cause incalculable damage. Some of the most destructive diseases of the cereal crops are caused by the ravages of microscopic fungi, which attack respectively the flower, the grain, the leaves, the chaff, and the straw. Those who have seen corn fields in July, when the flower is bursting through the sheath, must have often noticed several greyish-black heads appearing here and there among the verdant stalks. In some fields these are few and far between; in others they are more numerous, almost every

alternate stalk presenting this amorphous appearance. When one of the heads thus affected is

pulled and examined,
every chaffy scale is
found to be filled
with a firm black
matter, like soot ag-
glutinated by mois-
ture. This strange
phenomenon is attri-
buted to the state of
the air, to the condi-
tion of the seed, or to
the character of the
soil ; but there are
few comparatively

FIG. 44.—USTILAGO CARBO.   who are aware of its

vegetable origin,—who know that it is owing to
the development of minute parasitic fungi, favoured

of course by unhealthy conditions of the atmosphere and the
soil. To botanists it is known
under the name of *Ustilago carbo*
(Fig. 44), and by farmers it is familiarly called smut or dust-brand.
It is more frequent in corn than
in any other of the cereal crops.
Examined under the microscope

FIG. 45.—TILLETIA
CARIES.

Spores and Mycelium
highly magnified.

each grain is found to be converted into a vast
number of minute round balls or sporules of a
deep brownish-black colour. Bauer says that in
the 160,000th part of a square inch he counted
forty-nine of those sporules, so that eight millions
of them will exist on one square inch of surface.
Farmers regard the appearance of a few such
diseased ears among their corn fields with complacency—imagining that somehow or other they
are the harbingers of a good crop. There have
been frequent coincidences of this kind no doubt ;
but there is no connexion between them as cause
and effect. Appearing so early in the season, the
smut ripens and scatters its seed long before the
grain reaches maturity ; and by the time of harvest
not a trace of its existence remains to remind the
farmer of the ravages it has produced. This disappearance of the fungus when the crop is reaped,
especially if the harvest be good, is probably the
true reason why the farmer is prepossessed in its
favour. Were he better acquainted with its nature
and habits, he would look upon each black head
of corn with dread as the advanced guard of an
immense army of destroyers, lying in ambush in
the soil and on the seed which he sows, and ready
to take advantage of every favourable opportunity
to dash his hopes to the ground.

On the grains of wheat an equally common but

still more injurious fungus is developed called
bunt (*Tilletia caries,*
Fig. 45). In this
disease the seeds re-
tain their original
form and appearance;
indeed, the bunted
grain is plumper and
of a brighter green
than the rest, but the
inside is completely
converted into one
mass of black dust
of an exceedingly
fetid odour, and
greasy to the touch.
The surface of the
spores is beautifully reticulated. When the wheat
is thrashed, the infected grains, which retain their
original shape to the last, are crushed, and the
spores, being thus dispersed, adhere to the sound
grains by means of the oily matter which they
contain. Bunted wheat contains an acrid oil,
putrid gluten, charcoal, phosphoric acid, phosphate
of ammonia and magnesia, but no traces of starch,
the essential ingredient in human food. When
the black powder is accidentally mixed with the
flour, it gives it an exceedingly disagreeable taste,

FIG. 46.—PUCCINIA GRAMINIS.
(*a*) Slightly magnified. (*b*) Sporidia, highly
magnified.

and is probably injurious to health, though this has not been clearly determined. I have found a few sporules at the apex of every grain in very fine samples of wheat. Like Cadmus, therefore, the unconscious farmer sows in his fields with his seed the teeth of a dragon, which, if developed by a cold and wet season, will destroy his prospects of a harvest.

Every farmer has painful· knowledge of the disease called mildew (*Puccinia Graminis*, Fig. 46). It attacks the leaves and culms of corn, as well as many of the grasses employed in hay-making, and proves most injurious when developed to a great extent, as is often the case when severe frost immediately succeeds copious and continuous rain in autumn. It appears on the diseased leaves in yellowish elongated pustules, which speedily burst into fissures filled with an orange powder, diffuse themselves, and become confluent, until the whole plant looks as if sprinkled with red ochre. These rusty spots, under the microscope, are found to consist of a number of filaments aggregated together, on which are developed a cluster of colourless globose cells, which, as they enlarge, become filled with an orange-coloured endochrome. A month or two later the red pustules assume a browner hue ; and if examined under the microscope they are found to consist of a dense

aggregation of club-shaped bodies, their thicker
end being divided into two chambers, each filled
with minute sporules, and
their lower end tapering
into a fine stalk connecting them with the stem
of the corn (Fig. 46, *b*).
These different forms were
supposed to be distinct
species, but they are now
ascertained to be dimor-
phic forms of the same fungus, produced by the
modifications of growth.   Attention has already
been directed to the relationship between this
species and the Berberry-rust, which is supposed
to be only another form of it.   When this disease
is very prevalent, the juices of the corn are inter-
cepted, the stimulating effects of light and air are
prevented, and the grain in consequence becomes
shrivelled and defective, yielding at the same time
a superabundant quantity of inferior bran.

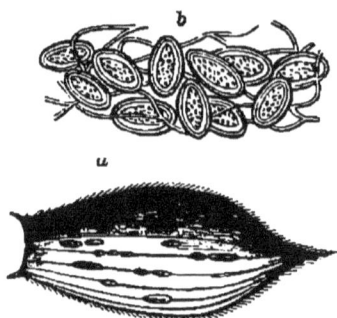

Fig. 47.—Puccinia Graminis.
(*a*) Diseased chaff-scale. (*b*) Spores.

One of the most remarkable diseases affecting
the cereals is ergot, already slightly alluded to.
Though found in various kinds of grasses, such as
Agrostis, Festuca, Elymus, and Dactylis, this
disease is most frequently produced in rye, and
hence it is commonly known as ergot of rye.   It
is not very common, although diffused in greater

or less abundance throughout the whole of Great
Britain; but in the zone where rye is the prevailing
grain, comprehending all the countries bordering
on the Baltic, the north of Germany, and part of
Siberia, it occurs in great abundance, and is often
a cause of much distress. It is owing to the growth
of a fungus called *Claviceps purpurea* (Fig. 48),

which converts the ovary of the
grain into an elongated cylindrical
excrescence, a little curved, and
somewhat resembling a horn or spur
projecting from the chaff, and hence
the rye thus affected is called in
common language spurred-rye. The
grain when attacked becomes first
soft and pulpy, afterwards it hardens and elongates gradually. It is

FIG. 48.—CLAVICEPS
PURPUREA.

first of a red or violet colour, afterwards lead-
coloured, and finally black with a white interior.
It contains so large a proportion of oily inflammable matter, that it will burn like an almond when
lighted at a candle. Generally only two or three
grains in a spike are affected, whose nutritious
part is thus completely destroyed, and converted
into a highly injurious substance. When rye is
extensively cultivated, grains diseased in this way
often compose a considerable part of the bread
produced, and thus not unfrequently give rise to

ergotism, one of the most distressing diseases
with which the human frame is affected. Professor Henslow, by way of experiment, gave it to
various domestic animals, mixed with their food,
when it was invariably found to produce sickness,
gangrene, and inflammatory action so intense, that
the flesh of the extremities actually sloughed away.
It is not, therefore, unlikely to have been the un-
suspected source of several strange morbid dis-
orders which have prevailed from time to time
among the poor in those places where rye is the
staple grain, and which have proved so perplexing
to the physician.   Professor Henslow published
a series of remarkable extracts from the parish
register of Wattisham, in Suffolk, in the year
1762, recording the sufferings of several persons
from an unusual kind of mortification of the limbs,
which was produced, in all likelihood, by the use
of spurred rye as food.   In some districts in
France, gangrenous epidemics, accompanied by
the most dreadful symptoms, used to be very prevalent in certain seasons ; but owing to the pains
taken to prevent ergot being sent to the mill and
ground up with the flour, they are now almost unknown.   Sheep and cattle allowed to browse in
meadows where ergot exists not unfrequently slip
their young, and become violently ill ; and pigs,
running about certain lanes and hedgerows where

the fungus often lurks in the shaded grasses, become diseased. Some places are so notorious for casualties of this kind connected with them, whose cause is not suspected, that owners of animals are afraid to allow them to be at large. The necessity of carefully picking out the ergoted grains whenever they are perceived in samples of wheat, cannot be too strongly or frequently impressed upon the farmer ; and wherever gangrenous diseases or uterine derangements prevail, search should be made for it in the neighbourhood, with a view to prevention. This curious disease, upon which more has been written by medical and botanical authors than upon almost any other vegetable production, affords one of the most extraordinary examples, within the whole range of physiology, of a natural chemical transmutation ; the nutritious grain being metamorphosed, by the agency of a fungus, into a hard horny substance, endowed with properties the very reverse of its original wholesomeness, and ministering suffering and death instead of life and strength to those who partake of it. Strange to say, however, the children in some parts of the north of Europe eat with impunity immense quantities of this diseased rye, under the name of St. John's Bread. This is an extraordinary instance of the uncertain effects of the same species of fungi upon the human constitution, and the wide differences they exhibit in their qualities in different countries.

The first failure of the potato crop, which came like one of those sudden hurricanes of the tropics, carrying death and desolation in their train, is doubtless vivid in the recollection of all middle-aged people. This root, from its extraordinary productiveness, with little labour or exertion of any kind, became gradually a substitute in whole districts, especially in Ireland and the Highlands of Scotland, for the older cereal crops, as the staple food of the people ; so that when a blight fell upon it, and the crop everywhere completely failed, hundreds of thousands were deprived of their sole means of subsistence, famine and its consequent malignant fevers rapidly spread throughout the land, and the social and agricultural system based upon this uncertain and narrow foundation was convulsed and completely broken up. Nor was this disease a temporary scourge ; it has returned every year since, with more or less fatality, so that the potato has become one of the most troublesome and precarious of all our crops. The year before last the potatoes were irredeemably bad, having failed to the extent of three-fourths of the yield. A minute mould, called *Peronospora infestans* (Fig. 49), consisting of grey interwoven filaments, bearing stalks some of which have a swollen or moniliform appearance, probably owing to repeated abortive attempts to produce fruit, and others develop a number of ovoid and transparent spores, and among them bodies of a larger size which mature within themselves zoospores by a differentiation of their contents, is invariably connected with the disease, and is found on the decaying plants ; the growth of the fungus being aided by some predisposition

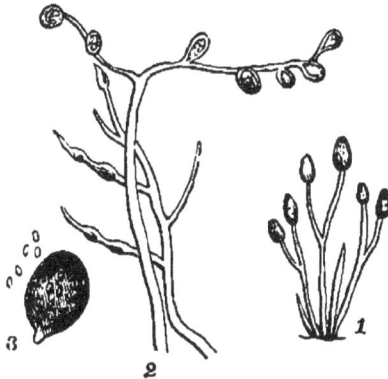

FIG. 49.—PERONOSPORA INFESTANS.
(1) Young plants. (2) Full grown. (3) Spore.—All magnified.

in the state of the vegetable, induced by the soil or the atmosphere. The potato is commonly affected after the tubers have been formed, and have attained a considerable size. It first attacks the leaves entering by the stomata or breathing-pores on the under side, and covering them with brown blotches, as if they had been burnt by the action of sulphuric acid. From the leaves it

spreads rapidly down the stem, till in a very
short time the whole of the plant above ground
is destroyed. The disease still spreads its
ravages, until ultimately it reaches and penetrates
with its spawn the tubers, the substance of which,
when affected, speedily turns brown, emits a
very peculiar and unpleasant odour, and soon
decays to a fetid watery matter. The diseased
tuber is strongly alkaline, from the presence
of ammonia ; the sound potato having an acid
reaction. The starch of the diseased potatoes
is often unaffected by the parasite, retaining
its nutritive properties, and may be separated
easily by rasping the peeled tubers upon a grater
into a tub of cold water.

The potato-blight belongs to a large genus of very
destructive fungi, affecting many of our vegetables
and fruits ; but as a species it is a comparatively
recent introduction. Facts derived from numerous sources lead to the conclusion that it did
not exist in this country previous to the autumn
of 1844. All the naturalists who examined it then
declared it to be quite new to them. It is considered by the most eminent botanists to be of
American origin, peculiar to the potato, and accompanying it wherever it grows wild in its native
country, as the smut accompanies the corn in this.
From South America it was first brought to St.

Helena by the north-east trade-winds, which bring
from the same continent those singular red dust
clouds which the microscope of Ehrenberg found
to be composed of vegetable organisms, and which
have served as tallies upon the viewless winds,
indicating the course of their currents. St. Helena
lies in the same latitude with Peru, and is nearer
the native habitat of the potato than any other
country in which the disease has been subsequently experienced. In this island, finding the
conditions of moisture and temperature favourable
to its development, it increased with amazing rapidity, loading the air with myriads of its impalpable seeds. Thence it seems to have been carried
by the winds to Madeira and North America ; and
so has progressed from country to country, gain-
ing new accessions of strength and numbers from
every field, speedily making its dread presence
known wherever it alighted. It reached England
in the autumn of 1844, and seems at first to have
been confined exclusively to the south-western
districts. From Kent it travelled west and north,
devastating the whole of Ireland and halting mid-
way in the South of Scotland, so that the crops
in the Highlands were that year free from the
pest. In 1846 it proceeded throughout the north
of Scotland, where its effects in certain districts
were scarcely less disastrous ; thence on to the

Shetland and Faroe Islands, and to northern
latitudes, as far as the limits of the cultivation
of the potato in that direction extended. On
the Continent it has been observed to progress in a similar manner; its geographical limits,
as well as its intensity, becoming more extended and marked with each succeeding year.
Why the fungus should have been introduced in
1845 and not in previous years, and why it should
then all at once have acquired such fearful power,
we cannot positively tell, any more than we can
tell why the memorable plague of London, or
those deadly pestilences which swept over Europe
in the middle ages, should have sprung up so suddenly as they did. Almost all authorities are
agreed that the disease generally makes its first
decided appearance during thundery weather;
and the exceptional amount of electrical disturbance which extended over the whole kingdom
during the summers of 1845 and 1872, seems to
have been most unfavourable to the potato-crop.
But these were not the only wet and thundery
summers that have been known in Britain since
the introduction of the potato; and the climatal
conditions of the various countries in which the disease appeared in 1845 were as different as possible.
It has therefore been suggested that cosmical
conditions may at definite intervals favour the

disease. History points out a singular correspond-
ence between the dates of several of the great
national epidemics of the middle ages and the ascertained period of maximum sun-spots, which is
between eleven and twelve years. And as this
period is accompanied by electrical and other dis-
turbances in the earth's atmosphere, it is highly
probable that it may be connected with the potato-
disease, which manifests a similar periodicity. The
year 1872 was near the maximum of sun-spots;
and that year witnessed the most virulent outbreak of the potato-disease since 1846; and just
as the epidemics of the middle ages were associated with cattle-disease and general failure of
crops, so the potato-murrain of 1872 was associated with epidemic diseases in animals and in
field and garden crops. That season was one of
the most unfavourable within this generation as
regards the fruits of the earth. No certain pre-
ventive of this destructive disease has yet been
discovered, notwithstanding the many plans proposed, which fail as often as they succeed.

I have said that the genus *Peronospora*, to
which the potato-parasite belongs, contains several
species which are often exceedingly destructive in
this country. They are the most common and
abundant of all fungi. For ages they have met
the eye in fields and gardens. Onions, cabbages,

turnips, beet-root, peas, gourds, spinach, tomatoes, and almost all the green crops we raise, often suffer severely from this blight.   In seasons favourable for their development they spread like wild-fire, and destroy everything before them.   A closely allied species is the cause of the destructive plague affecting the grape, which shortly after the potato-disease broke out, spread suddenly throughout the vine-growing countries of Europe.   The fungus called *Oidium Tuckeri* (Fig. 50), concerned in this epidemic, made its first appearance, or rather was first observed, in the hot-houses of Mr. Slater of

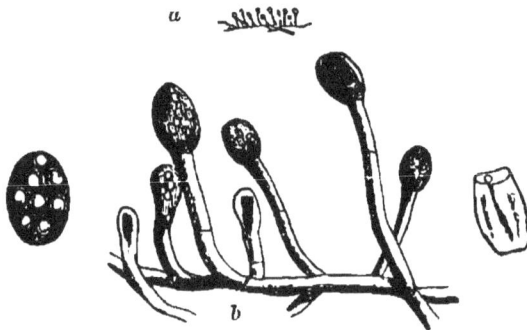

FIG. 50.—OIDIUM TUCKERI.
(*a*) Natural size.   (*b*) Magnified.

Margate by his very intelligent gardener, Edward Tucker, after whom, in consequence, it received its specific name.   It seems to have been previously unknown to botanists.   Its origin is very obscure. We know not whether the germs of the fungus spread from those produced in the hothouses of Margate, or whether similar conditions elsewhere prevailing originated it without any connexion existing between the places, but certain it is, that an immense profusion of the same fungus appeared almost simultaneously throughout the vineries in this country. Two years afterwards, the seeds borne across the Channel by winds reached France, where for a time their ravages were limited to the forcing-houses and trellised vines of Versailles, and several private establishments in the neighbourhood of Paris. But in 1851 it unhappily reached the open vineyards in the south and south-east of France, where it destroyed nearly the whole of the crop, rendering the grapes unfit for food or wine. With resistless speed it forced its way into the finest provinces of Spain, where so deplorably were the vineyards blighted by it, that in many places they were abandoned in despair. It crossed the Mediterranean to Algeria, extended its flight to the vine-clad slopes of Lebanon, ruined the currants of the Greek Islands and the raisins of Malaga, and destroyed so utterly the far-famed vintage of Madeira, that this wine for many years was numbered among the things that were. After raging for a number of years with similar if not increased violence, it subsided, like the potato-disease, to a certain extent,—whether owing to the remedies applied proving successful, or the conditions for its development proving unfavourable, it is impossible to say. Some places now enjoy complete immunity from it ; and in other places the cultivation of the vine, formerly abandoned, is resumed with vigour, and with every prospect of success. A large percentage of the crop is, however, season after season, still lost from this cause ; and probably the disease is now so completely established that it is vain to hope for its speedy disappearance.

The fungus which causes the vine epidemic is very minute, covering the affected grape like a white cobweb. From its radiating filaments several jointed stalks rise vertically like the pile of velvet, the upper joints swelling, assuming an egg-shape, and giving birth to the reproductive spores. It has been proved to be only a stage or incomplete state of an *Erysiphe*. It makes its appearance first as a minute speck on the grape when about the size of a pea. It speedily enlarges and covers the entire surface of the berry, investing it with a net-work of interlacing fibres, exhausting its superficial juices, and crushing it within its embrace. So richly is it furnished with the means of propagation, that a succession of seeds is developed by the same filament, and three or four ripen and are dispersed at the same moment ; while, so loosely are they attached to their receptacles, that the smallest breath of air or the least brush of an insect's wing carries them off to other grapes, to infect these with a similar blight.

I may remark here by way of parenthesis, that fungi have a special and inordinate predilection for the produce of the vine in all the stages of its history and manufacture. One species, as we have seen, luxuriates on the grape; another fungoid form is concerned in the process of fermentation, and the consequent resolution of the grape juice into an alcoholic product; a third, as already mentioned, frequents, like a Bacchic gnome or convivial Guy Fawkes, the vaults where wine is stored up, forming a most remarkable and picturesque feature in that vast temple of Silenus— the London Docks—hanging down in immense festoons from the roof of the crypt, swaying and wavering with the least motion of the air like dingy cobwebs. Private cellars are not unfrequently drained dry by a host of thirsty vegetable topers in the shape of huge fleshy fungi, developed by the moist, dark atmosphere of the place, and the rich pabulum of saccharine food which they find there. The bottle of port brought up to table, whose venerable appearance the host eyes affectionately, and the guest with eager expectation, sometimes affords a melancholy illustration of the vanity of earthly hopes. A cunning fungus has been beforehand with them; and like the famous rat, whose inventive powers were quickened by necessity, which drew up the liquid contained in a bottle by dipping its tail into it, the vegetable, equally sagacious, develops itself first on the cork, and having penetrated it with its spawn, sends down long root-like appendages into the liquor, exhausting it of its rich aroma, and rendering it a mere *caput mortuum*. Nor is the wine left unmolested, even when it has been drawn into the decanter; a meddling fungus still follows it, and renders it sometimes mothery, the cloudy filamentous dregs left at the bottom indicating its presence. In short, in some shape or other this fungoid vegetation perseveringly accompanies the fruit of the vine in all its changes and transitions from the German hills to the British dining-room; and like an ill-odoured exciseman, levies a tax upon it for the benefit of its own constitution.

A very familiar example of an Oidium occurs on decaying oranges, commencing at first in minute pulverulent spots, which speedily become confluent, and of a deep greenish-grey tinge. This genus of fungi is very destructive to fruits of all kinds. The lover of fruit may have often noticed thin concentric, cream-coloured, or fawn-coloured patches on the skin of apples, pears, and plums, producing very rapid decay. These patches are caused by *Oidium fructigenum,* which, when it has once obtained possession of a tree, spreads with fatal rapidity, destroying the fruit while still hanging on the branches. The *blanc de rosier* or *rose-blight,* which gives a melancholy leprous appearance to the leaves and calyces of roses, used to be called *Oidium leucoconium;* but this stage was only an incomplete or conidiferous condition of *Sphærotheca pannosa.* An allied species constitutes the hop-mildew, which has often proved so disastrous in the hop-gardens of Kent, and raged as an epidemic in 1854.

All the mildews and blights hitherto described are light-coloured; but there is another class of fungi, equally destructive, called black mildews. They are caused principally by species of Antennaria and allied genera, which form thick, black, felt-like patches on leaves, disfiguring trees, and injuring them fatally, by closing up their pores, and preventing the free admission of the air; and also by depriving them of the full, direct light of the sun. They are principally developed on those leaves which had previously been covered with the honey-dew of aphides or plant-lice. Whole plantations are often literally covered with sackcloth and ashes. In the Azores the orange-groves of St. Michael have suffered dreadfully from this cause ; while in Ceylon the coffee-plantations, in the south of Europe the olive-trees, and in Syria and China the mulberry-trees, have sustained of late years immense damage from an unusual development of black mildews. Few objects, it may be remarked, are more beautiful under the microscope than the wheel-shaped, ray-like processes which radiate from the seed-bearing organs. A species of this family, called *Fusarium mori*, is produced in such abundance on the leaves of the mulberry, in Syria and China, as materially to diminish the supply of food provided for the silkworm.

But it is not only in food and luxuries that man suffers from the ravages of fungi ; he also suffers in his property. Builders have painful knowledge of one or two species, known under the common name of dry-rot. This most destructive plague is usually caused in this country by the *Merulius lachrymans* (Fig. 51). It occurs on the inside of wainscoting, in the hollow trunks of trees, in the timber of ships, and in the floors and beams of buildings in moist, warm situations, where there is not a free circulation of air. It appears at first in round, white, cottony patches, from one to eight inches broad, which afterwards develop over their whole surface a number of fine yellow, orange, or reddish-brown irregular folds, most frequently so arranged as to have the appearance of pores, and distilling drops of. moisture when perfect ; whence its specific name. In the mature state it produces an immense number of minute rusty sporules, which alight and speedily vegetate in the circumjacent timber, destroying its elasticity and toughness, and rendering it incapable of resisting any pressure, until gradually it crumbles into dry

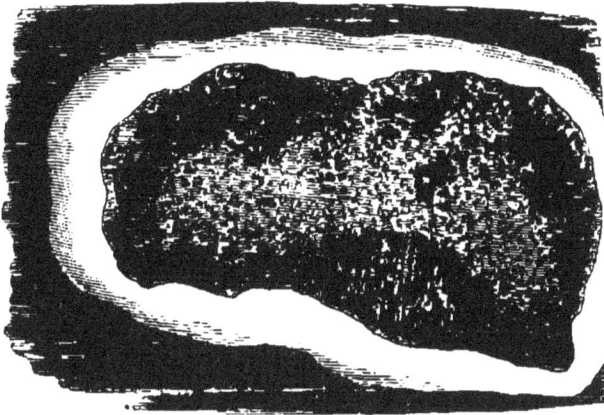

FIG. 51.—MERULIUS LACHRYMANS.

brown dust. This insidious disease, once established, spreads with amazing rapidity, destroying some of the best and most solid-looking houses in a few years. The ships in the Crimea suffered more from this cause than from the ravages of fire, or the shot and shells of the enemy. The dry-rot of oak-built vessels is caused, however, not by the *Merulius* but by the *Polyporus hybridus*, whose

2 D

mycelium forms a dense membrane or branched creeping strings, while the pores of the hymenium are long, slender, and minute. So virulent is the nature of dry-rot that it extends from the wood-work of a house even to the walls themselves, and by penetrating their interstices, crumbles them into pieces. We have every reason to believe that the leprosy of houses, so graphically described in Leviticus, was a dry-rot caused by some species of Merulius or Polyporus ; the materials and sanitary condition of Eastern houses being peculiarly favour-able for the development and spread of fungous growth. ' I knew,' says Professor Burnett, ' a house into which the rot gained admittance, and which, during the four years we rented it, had the parlours twice wainscoted, and a new flight of stairs, the dry-rot having rendered it unsafe to go from the ground floor to the bed-rooms. Every precaution was taken to remove the decaying timbers when the new work was done ; yet the dry-rot so rapidly gained strength that the house was ultimately pulled down. Some of my books which suffered least, and which I still retain, bear mournful impressions of its ruthless hand ; others were so much affected that the leaves resembled tinder, and when the volumes were opened, fell out in dust or fragments.' Many practical persons have written upon this disease ; and

the remedies proposed are as numerous as
their authors. But the only certain preventives
of the evil seem to be the removal of the decaying and contagious matter, the kyanizing or
impregnating of the surrounding wood with a
strong solution of corrosive sublimate or coal-tar
and the admission of a free current of air. Much
also may be done by cutting timber destined for
building purposes in winter, when fungi are
usually dormant or dead, and properly seasoning
it by steeping it in water for some time, and then
thoroughly drying it before it is used. Houses,
in order to be free from this plague, should be
built in dry, open, and airy situations, and
efficiently ventilated throughout every part, especially of the wood-work ; when these conditions are
observed, this evil will disappear.

The rapidity with which the spawn of fungi will
spread, and the depth to which it will penetrate, are
truly wonderful. The most solid timber, in a few
months, when exposed to the weather in favourable
circumstances, will often show traces of spawn.
Elm trunks when felled quite sound, by the
second year are penetrated to the very core with
it. The beautiful and well-known Tunbridge-
ware is formed of ordinary British oak, coloured a
rich mineral-green by the presence of a little
fungus (*Helotium æruginosum*), whose spawn, when

a portion is examined under the microscope, is seen to traverse the whole fabric of the wood with a beautiful net-work of minute threads, giving it the peculiar green tint which is so much admired. The fungus attacks fallen oak branches; but while it permeates the wood with its spawn, it seldom produces the little green open cups of its mature state, so that the origin of green wood is mysterious to the uninitiated. Of course, such timber being in a state of decay is worthless except for ornamental purposes. Gardeners frequently complain of the spawn of fungi destroying trees and herbaceous plants. Rhododendrons in shrubberies are often rendered sickly by a substance resembling sarsaparilla, or a mass of branched and corrugated roots about the thickness of a tobacco-pipe, traversing the soil and twisting round the roots of the shrubs an inch or two beneath the surface. This parasite is supposed to be an anomalous condition of some fleshy fungus, such as *Agaricus grammocephalus*, which is common in shrubberies, sending out its branched root-like strings in all directions to a great distance. Whenever the parasite is removed the shrubs recover their proper healthy character. Old hidden stumps and decayed roots, owing to the spawn of fungi which they contain, are peculiarly dangerous to young plantations. It is not always safe, for the same reason, to plant new trees in situations where old trees have drooped and perished. Compost, consisting of decayed vegetable substances, and the fallen leaves and twigs that strew the ground in shrubberies and plantations, by harbouring the spawn of fungi, may prove the source of serious mischief; and many diseases of cultivated crops may be traced to the existence of spawn in the manure. To the farmer and gardener, therefore, the nature and effects of fungi should be subjects of special study, seeing that they are so powerful for evil.

In concluding this notice of the destructive fungi, mention may be made of a peculiar form of Penicillium or mould, which is almost invariably present in the solution of copper employed in the process of electrotyping. It proves an intolerable nuisance, inasmuch as it is often invested with a silver coat, and injures the beauty and the finish of the articles which are subjected to the process. It is extraordinary that the poisonous nature of the solution does not destroy it; but it has been often observed that various species of mould luxuriate in solutions of arsenic, opium, and other poisonous chemical substances, which would prove instantly fatal to all other plants. The factories of India occasionally suffer greatly from the presence of fungi on extracted opium.

After this detailed description of the specific

fungi connected with the more remarkable kinds of vegetable epidemics, a few words regarding their mode of increase and dispersion may not be uninteresting. It is a well-known physiological axiom, that the simpler an organism, the more bountifully is it furnished with the means of propagating itself. Exposed to numerous contingencies, failure of reproduction by one method is compensated by the development of another. Accordingly, fungi are provided with two, three, and in some cases even with four, modifications of reproductive power, all equally effectual, though not all developed at one and the same time. Every filament or cell may contain its germs, and each germ spring up into new forms equally fitted for propagation in the space of a few hours ; nay, some may pass through the course of their existence in a few minutes, and give birth to thousands even under the field of the microscope. In truth, the common reproductive bodies called spores do not directly propagate the fungus. They germinate, however, at definite points, and after a time produce on their filaments secondary and even tertiary spores, which are the true organs of reproduction. These, by their minuter size and the more delicate spawn which they produce when germinating, enter the breathing-pores of the stem and leaves, and the tender tissue of the spongiolets of the roots of the wheat and corn, where the primitive spores and the large blunt spawn which they form could find no lodgment. In addition to the spores or conidia of the potato-blight, the fungus produces also zoospores or moving spores, furnished with the well-known moveable cilia, and capable of germinating and penetrating the tissues of the potato like the ordinary spores (Fig. 52). In the dewy autumnal mornings, when the potato-leaves are all moist, a few infected plants, by the aid of these swimming spores, will rapidly infect a whole field; while there is reason to believe that, in the form of oospores or sclerotia, the fungus hybernates during the nine months of its disappearance every year, and prepares itself for its annual attack in autumn. Furnished with such powers as these, it is a fortunate circumstance that the fungi connected with vegetable epidemics require peculiar atmospheric and other conditions for their growth, and when these are absent they will not develop themselves or spread; otherwise the whole world would be speedily overrun with them, and the fig-tree would not blossom, and there would be no fruit in the vines, the blossom of the olive would fail, and the fields would yield no meat.

It is worthy of remark that the destructive effects of all these parasitic fungi may, in most circumstances, be easily neutralized or prevented by a little intelligent forethought, care, and industry ; and providing incentives as they do to the exercise of these qualities, they compensate morally in some measure for the physical evils

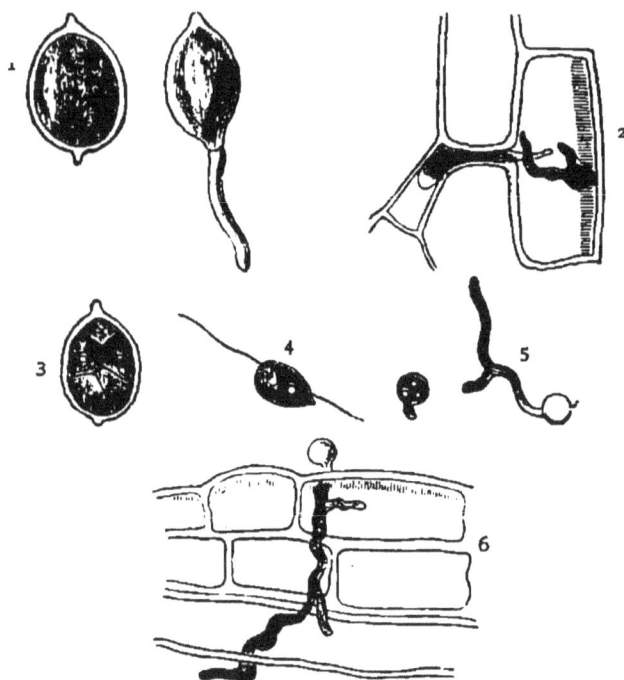

FIG. 52.

(1) Spores or conidia of *Botrytis infestans* germinating.   (2) The same sown artificially, and penetrating the tissues after eighteen hours.   (3) Spore with contents differentiated.   (4) Zoospore.   (5) Zoospores germinating. (6) Zoospore sown artificially on the stem, and after twenty-four hours penetrating the tissues and entering the intercellular spaces.

they occasion.   Certain conditions, as I have said, are necessary for their development, and it is to obviating and removing these that the builder and the farmer must look for exemption from the destructive vegetable diseases that affect their properties. It has been ascertained, for instance, that rust and blight arise from the over-manuring of fields ; the grain gorged with too copious a supply of nutritious juices, being brought into a favourable condition for the development of the dormant seeds of fungi which the wind may have wafted to it. The tendency in corn to form these diseases, therefore, may be prevented by moderate manuring, or by a free use of saline manures ; while the seed before being sown should be steeped in a corrosive solution or in brine. With regard to mildew in wheat, it has been suggested by Mr. Tycho Wing, as a remedy, that no reeds or loose grass should be allowed to remain in the ditches, but everything should be cleared away, and consumed at once. 'As the species which attacks reeds and grass is to all appearance the same with that of the wheat, the disease may be propagated in the spring from such outliers. For the same reason, it is desirable that the stubble should not be left on the land too long, and, indeed, long mowing must be better than reaping.' The various mildews that appear on the grape and other fruits and useful plants, may be prevented from developing themselves by the application, at an early stage, of powdered sulphur, which, combining as it does with the oxygen

of the atmosphere, forms sulphuric acid, the only
chemical poison destructive to moulds and mildews.

Fungi, owing to their cellular and perishable
nature, do not usually occur in a fossil state.
Some slight traces of them, however, now and
then occur among the relics of a former state of
things. Species of mould have occasionally been
found in the amber beds of the tertiary formation—
having been deposited and developed on the re-
sinous juices of the amber pines, just as filaments
of mould are often seen at the present day adhering to the gum of apricot and cherry trees. These
tiny plants, identical as they are with the common
green and blue moulds that infest our cupboards,
leave us no room to doubt that fungi were as prevalent and destructive in former epochs as they are
now. M. Goeppert, who has examined minutely
the amber of various lands, has detected in it,
besides moulds, fragments of mosses, hepaticæ,
and lichens, perfectly preserved, as in a mummy
case,—the sole insignificant relics of that vast
array of cryptogamic plants, which helped to
preserve the balance between the kingdoms of
the ancient world. Nature by this curious process of embalming, has perpetuated that which
a breath of wind was sufficient to destroy, and
"moulded into a geologic specimen what a

finger's touch would fade." While rocks and forests have been destroyed, without leaving a trace of their existence behind, the most delicate and fugacious organisms have been handed down to us in the most beautiful preservation from the remote postpliocene period—the temporal and fragile thus transformed into the eternal.

Having thus given a somewhat lengthened and detailed account of the structure, properties, uses, and other peculiarities of this curious and interesting tribe of plants, it may be proper, in conclusion, to glance at the place which they occupy as æsthetic objects in this fair creation. The careful observer will find the universal spirit of beauty sometimes as richly represented in these productions of corruption and decay, as in the more admired products of the vegetable kingdom. The very commonest fungi, which grow in the darkest and dreariest spots, are invested with a beauty, not essential to the part which they perform in the operations of nature, or to the efficiency of the organs, whether of absorption or reproduction, with which they are furnished. The fructification of one is a most graceful umbrella, adorned with delicately-shaded hues, and with exquisitely carved veils, fringes, and gills ; that of another presents the most beautifully sculptured ivory pores and sinuosities or richly-coloured tubes or spikes.

One species looks like a ruby cup; another is embossed with stars; while the leaves and the grasses of the woods and fields often form niduses for some of the loveliest and strangest forms, which our great Creator has scattered over the earth with lavish hand to delight the intelligent eye. Mrs. Hussey in this country has illustrated the higher tribes with a beauty and grace which leave nothing to be desired; while the Memoirs of lower species by the brothers Tulasne in France are adorned with the most exquisite engravings which art has ever produced. No natural objects lend themselves so appropriately for purposes of pictorial embellishment as fungi; while, with the exercise of a little taste in grouping, they might be made exceedingly effective at flower-shows, and in the adornment of the drawing-room in the appropriate season. Some of the most picturesque and attractive forms in the whole vegetable kingdom are found in the genus *Geaster* resembling star-fishes and sea-urchins ; while the *Aseroë rubra*, with its bright forked rays of a bright scarlet above, and pale rose below, and its large central aperture in the disk, looks lovelier than any sea-anemone.

There is not in nature a more picturesque object to the painter, or a more interesting study to the botanist, than the old decaying stump of a tree in some lonely haunt of a shady ancestral wood, where the soil, enriched by the organic contributions of centuries, is bursting into life through every crevice and on every inch. Such a stump, as Wordsworth beautifully says of the mountain, is 'familiar with forgotten years.' It is long since the tall massive oak which it supported has been removed by the axe, leaving a gap which the encroaching trees around strive in vain to conceal ; and nature has kindly smoothed away the traces of man's harsh treatment, and brought it back to perish on its own bosom. Every sunbeam and rain-drop that descended upon it, while crumbling it more, increased its picturesqueness, and while depriving it of its own life, helped to develop upon it other forms of life lower in the scale, until now, it not only adds to the air of antique mystery which pervades the scene, but peoples it with all the fantastic tenantry of Shakespeare's fairy land. In one corner may be observed a cluster of elegant pearl-like mushrooms, wee elfin-looking things with long, black stalks, and white wheel-like heads ; in another, the corky leaves of a *Thele-phora* closely pressed to the wood, with shell-like patterns, and colours as beautifully and dimly shaded on its surface as in a misty rainbow ; here the soft, viscid, flesh-like knobs of the *Tremella sarcoides*, resembling tiny teats,—or the

wrinkled, quaking, gelatinous mass of the witches'
butter, looking more like a frothy exudation from
the stump itself than a plant ; there a *Spathularia*
protruded from a wide mouth-like gap, like an
old woman's tongue, frightening away every
young rustic, full of the adventures and trans-
formations of the Seven Champions of Christen-
dom, from plucking it off, lest the owner, a metamorphosed witch perhaps, should return in proper
person to demand her unruly member, and inflict
a proportionate punishment.   In the middle of the
squared top, covered with the minute scurf-like
germs of unknown plants, are clustered the beautiful round vermilion balls of the *Lycogala*, or wolf's
milk, which, when bruised, exude a dark, grumous
liquor like clotted blood ; while springing from the
crevices of the bark, near the ground, the *Agaricus
necator* overtops the rest, with its zoned and olive-
coloured cap and dusky stem, distilling, when
broken or injured, a blood-like fluid, as though it
were a sensitive creature, or the sanguine *Stereum*
creeps by its side, bleeding when wounded, thus reminding one of Dante's terrific picture of the living
Avernus forest, where one feared to break the
boughs lest they should cry to him from the rents.
All these, with a score of other curious microscopic plants, hiding themselves from the superficial observer, but revealing themselves openly

to a loving scrutiny,—cup-lichens and trailing
green mosses, and slimy green dust-like confervæ, surrounded perhaps with a border of dock-
leaves, or a fringe of palmy ferns,—invest the aged
stump with a nameless charm in the estimation of
all true lovers of the picturesque. In such a place
one realizes the vividness of Shelley's description
of the garden of 'the sensitive plant :'

> 'And agarics and fungi, with mildew and mould,
> Started like mist from the wet ground cold ;
> Pale, fleshy, as if the decaying dead
> With a spirit of growth had been animated.'

And returning from the woods and the fields to
the retirement of our own homes, we find that
there are forms to be seen there as beautiful and
suggestive of curious thought, as any we have seen
in the wider field of nature out of doors. If we
examine under the microscope the green or grey
covering which spreads over damp walls, or enve-
lopes a stale piece of bread or fruit in a cupboard,
or creams over the surface of preserves, what a
wonderful scene of beauty, a delicate soft world
of white, an in and out of living lace, suddenly unfolds itself like a miracle to our view ! Thousands
of plumy trees and feathery fern-like plants rear
themselves up in every conceivable attitude, and
all so tender and transparent that the minute seeds
are seen lodged in the interior of their stems;

luxuriant forests draperied with pendent parasites, and milk-white mosses enveloping the ground, and clothing old, rotten-looking stumps with beauty, all busy in the fulfilment of their offices, lengthening and swelling, and falling, and scattering their minute seeds in little white clouds up and down upon the surrounding air. He who is privileged to feast his eyes on such a witching spectacle as this, must deeply feel that the Spirit of beauty is working everywhere, covering the obscurest nooks with the same glory as the grandest scenes, and using the same pencil to deck the humblest fungi that redeem the decay of the grave from its hopelessness, and to gild with living radiance the constellations that light up the spacious avenues between this world and the unseen.

INDEX OF SCIENTIFIC NAMES.

2 E

436INDEX.

INDEX OF POPULAR NAMES.

Edinburgh University Press: T. & A. Constable, Printers to the Queen.